AN IRISH HERBAL

An edited translation of *Botanalogia Universalis Hibernica*,
or A General Irish Herbal, which contains an alphabetical
listing of the plants, herbs, flowers, shrubs and trees of the
British Isles and the uses to which each may be put in the
formation of cures.

AN IRISH
HERBAL
The *Botanalogia Universalis Hibernica*

by
JOHN K'EOGH

Edited by
MICHAEL SCOTT

THE AQUARIAN PRESS
Wellingborough, Northamptonshire

First published 1735

This edition first published 1986

© MICHAEL SCOTT 1986

British Library Cataloguing in Publication Data

K'Eogh, John
An Irish herbal : Botanalogia Universalis Hibernica.
1. Plants, Useful — Great Britain — Dictionaries
I. Title
II. Scott, Michael
III. Botanalogia Universalis Hibernica. *English.*
581.6'1'0941   QK98.4.G7

ISBN 0–85030–532–2

The Aquarian Press is part of the Thorsons Publishing Group

*Printed and bound in Great Britain by
Whitstable Litho Ltd., Whitstable, Kent*

Contents

Acknowledgements

I would like to express my gratitude to William Holland for his invaluable assistance, both in the editing and preparation of this work.

I would like to thank Aidan Heavey, Padraic O Tailliuir and Gus MacAmhlaigh for kindly allowing me to work from their editions of *Botanalogia Universalis Hibernica*.

A special word of thanks to Andrew Waters.

For Anna.

The Author
of
Botanalogia Universalis Hibernica
John K'Eogh
1681? – 1754

The second of twenty-one children, John K'Eogh or Keogh followed his father into the Church and became chaplin to James King, fourth Lord Kingston, before finally obtaining the living of Mitchelstown, in Co Cork. He married Elizabeth Jennings, the daughter of Dr Henry Jennings, a cousin of the Duchess of Malborough.

John K'Eogh's first work, *Botanalogia Universalis Hibernica*, was published in Cork in 1735, and immediately established itself as an important record of Irish plants and their medicinal properties.

There is not an herb, shrub, or tree in nature but that it is servicible to man either for food or medicine, or for both.

John K'Eogh
Botanalogia Universalis Hibernica, 1735

Introduction

In this age of technological-assisted impersonal medicine, it is perhaps paradoxical that man has recently returned to an earlier form of healing — that of herbalism. Part of the attraction of herbal medicine and herbal cures is that cures are effected by wholly natural means, without artificial additives or synthetic drugs, and are readily and cheaply available — sometimes in our own gardens.

Herbalism and herbal and plant cures have, of course, a very distinguished history, and at one time every town and village had its herb doctor or wise woman dispensing simples, poultices and philtres. Unfortunately, the art of herbalism became tainted by its association with witchcraft and, indeed, in many people's minds the two were almost indistinguishable, and this led to its falling into disrepute and disuse.

But the healing properties of plants and herbs have always fascinated mankind, and there are European herbals in manuscript form dating back as far as the tenth century, while Middle Eastern and Oriental herbals existed five centuries prior to that. The great renaissance of herbals began in the last few years of the sixteenth century with the publication of *The Herbal or a General History of Plants*, which is more commonly called *Gerard's Herbal*, and which was first published in 1597. Fifty years later saw the publication of possibly the most famous herbal in the English language, Culpeper's *English Physician*, but which is now more usually known as *Culpeper's Herbal*, which first appeared in 1652. Indeed, both of these herbals have remained in print almost continuously since their first publication.

John K'Eogh's *General Irish Herbal* first appeared in

1735, and differs slightly from most of the other herbals in that it is solely concerned with the plants, flowers, herbs, bushes and trees of the British Isles. It is similar to Culpeper in that it provides a description of the plant, when and where it may be found, and then goes on to detail its virtues. The herbal differs from Culpeper in that it is a much more concise work. K'Eogh's herbal remained for many years one of the standard reference works on the virtues of the flora of the British Isles. However, copies of the original work are now extremely rare, and the work was sadly forgotten. This edited version is the first reprint since the first edition of 1735, and the cures remain as effective today as they were when the book was first published over two hundred and fifty years ago.

Preface

This undertaking of *Botanalogia Universalis Hibernica*, a general Irish herbal, was never attempted by anyone before, and so, gentle reader, if I fall short of it, I hope you will be so good as to excuse me, for it is a science that is chiefly built upon daily experience which in turn leads to daily improvement.

I am convinced that if the real properties, or true qualities of all the herbs and trees growing in this Kingdom were discovered, there is not a distemper or a disease which the inhabitants presently suffer from, which might not be cured. And I am referring to a single herb, tree or shrub which, if not completely curing, might, at least remove the ill for a considerable time, without having to send for exotic herbs and drugs from foreign countries, which often destroy more than they cure.

I am also of the opinion that these drugs which are made up or compounded in druggists or apothecaries shops have much the same ill effect. Moreover, in the said shops, there are often a great many decayed slops and drugs which they sell as good and fresh, and by which means how many patients have been destroyed? Therefore the best and safest method is to make use of simple herbs, the products of our own Kingdom, whose qualities and virtues are by long experience perfectly known to us.

John K'Eogh

A

ABELE TREE or
White Poplar
(*Populus alba*)

The timber of this tree is white, and not very hard to work. The bark is smooth and whitish, the leaves are round with pointed corners — white, smooth, soft and downy on one side and green on the other. It is planted about mansion houses for shelter, and it is of quick growth.

The juice of the leaves of this tree eases pains in the ears, and heals ulcers and eruptions on the skin. The bark is useful in promoting the discharge of urine and is therefore good against the strangury (a difficulty in discharging urine).

ADDER'S TONGUE
(*Ophioglossum sive lingua serpentina*)

This is a small tender plant, about 4 or 5 in high, consisting of a single, thick oval, smooth leaf, from the bottom of which rises a stalk about 2 in high, bearing on the top a slender crenated tongue. It grows in moist meadows and is in its prime in the month of May.

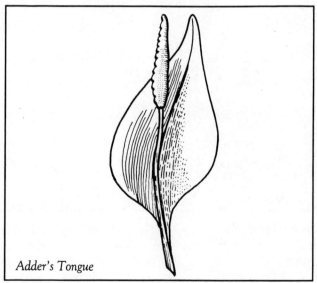

Adder's Tongue

Adder's Tongue is an excellent herb in the healing of wounds. Taken inwardly it is good against bruises, wounds and inflammation of the liver. An oil or unguent made from it can be applied to inflammations, burns, wounds, St Anthony's Fire and to all hot tumours.

AGRIMONY
(Agrimonia)

The leaves of this plant are long, hairy and serrated about the edges, almost like the leaves of hemp or strawberries. The stalk grows to about 2½ ft high, and it brings forth small yellow flowers in long spikes, which are then succeeded by rough little burrs. It grows in hedges and the borders of fields and flowers in June and July.

Agrimony opens obstructions of the liver and wonderfully strengthens it. It purifies the blood, and is good against the strangury, and pissing of blood, and its seeds taken in claret is singularly good against the bloody-flux. The leaves, powdered with hogs lard, heal old wounds when applied in a warm poultice. If the leaves are bruised and a plaster made with the yolk of an egg, flour and honey, they can be applied to a cut, or a gall after riding.

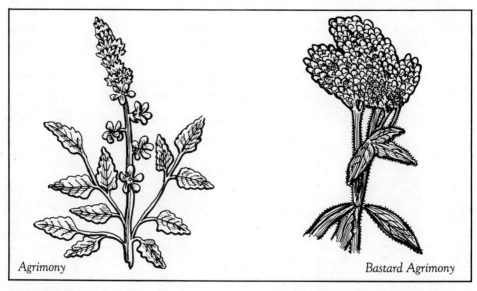

Agrimony

Bastard Agrimony

AGRIMONY, called Bastard or Water Hemp Agrimony
(Pseudohepatorium)

This has long round stalks full of white pith on which grow long blackish leaves, which are rough, hairy and serrated. On the tops of these stalks sprout many small flowers of a carnation colour. It grows in moist places near ditches and stagnated waters.

Agrimony is exceedingly good against opulations of the liver and spleen, and it is also useful against internal wounds and bruises, and if the leaves are bruised and mixed with the yolk of an egg they can be applied to a wound or cut.

ALDER or Aller Tree
(Alnus vulgaris)

This tree needs no description being so well known.

The bark or rind of it, because of its astringent quality is useful against the swellings of the throat. It heals and cauterizes sores and ulcers. It is often used by the common people as a black dye, and the leaves of it are made use of against ulcers and all kinds of inflammations.

ALEHOOF or Common Ground Ivy, or Gill Go by the Ground
(Hedera terrestris chamaecissus)

This plant has a strong earthy smell, and it grows everywhere in hedges and shady places, flowering in April. It needs no further introduction because it is so well known.

It is a great laxative, and can also cure deafness, coughs or any disorder of the lungs, it provokes urine and cleanses the ureters and, if it is steeped in brandy is of great service against the colic.

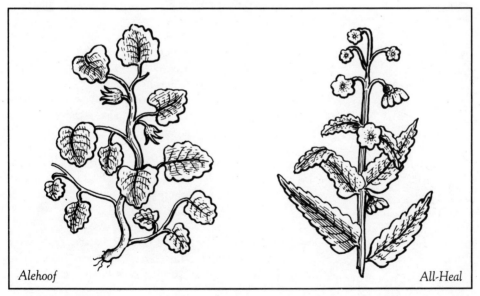

Alehoof

All-Heal

ALEXANDERS
(Hipposelinum seu smyrnium vulgare)

This bears large winged leaves of a yellowish-green colour, the stalks grow to be 3 or 4 ft high and on the tops of which grow pretty large umbles of small five-leaved white flowers. It grows upon rocks by the seaside, flowering in June, and is commonly preserved in gardens.

The leaves of this plant are commonly applied to wens and hard swellings to dissolve them. The seed is also used to remove all obstructions of the liver, spleen and kidney, and it provokes sweat, urine and helps to cure the jaundice and dropsy. It is commonly boiled and eaten with salt meat. The seed is mixed with tobacco and smoked in a pipe, the fumes helping to ease a toothache.

ALL-HEAL or Clowns All-Heal or Clowns Woundwort
(*Panax colono gerardi, Stachys palustris foetida*)

The stalks grow 2 or 3 ft high, are square and rough almost to pricklyness, the leaves are long, hairy, sharp pointed and indented. The flowers grow whorl-fashion towards the tops of the stalks, and are of a deep red colour. It grows in ditches and watery places, flowering in June and July.

All-Heal is an excellent herb in the treatment of wounds and is beaten into a poultice with hog's lard and applied to green wounds. It stops all sorts of haemorrhages. The decoction of the roots taken inwardly for a considerable time helps to cure tumours. The leaves, when stamped with an equal quantity of bay salt will cure the bite of a mad dog.

ALMOND TREE
(*Amygdalus*)

This tree is so like the peach tree with its leaves and blossoms that the only way to distinguish them apart is by their fruit. It is chiefly planted in gardens.

Bitter Almonds are used against all diseases of the lungs, liver and spleen and is therefore good against coughs, shortness of breath, inflammation and exulceration of the lungs. It should be taken in a sweet wine, and it is also an excellent cure against the headache when it is applied to the forehead with the oil of roses and vinegar. It is said that if a man takes five or six almonds before breaking his fast, then he will not become drunk that day.

Take 2 oz of the oil of sweet almonds, the same quantity of fresh butter, sugar candy and clarified honey, a quarter of grated nutmeg, which mixed together and taken off a liquorice stick, is an exceeding good cure for a cough.

HERB ALOE or Sea Houseleek
(*Aloes vulgaris sive sempervivum marinum*)

This is a purgative medicine and is frequently given to children for worms.

ANGELLICA or Garden Angellico or Wild Angellica
(Angellica saliva, Angellica sylvestris)

Wild Angellica is like that of the garden variety, only in that its leaves are not so deeply cut and that they are narrower and blacker, the stalks are much slenderer and shorter. It grows in shady places, near riversides, and in low-lying woods. The Garden Angellico flowers and seeds in June and July, and both varieties have much the same virtues.

The root, pulverized and taken inwardly causes sweat, expels all noxious humours and prevents any malignity that proceeds from the air; it also prevents the said malignity if it be chewed in the morning before breaking the fast.

A decoction of this herb helps to cure palpitations, oppressions of the heart and it also helps to promote urine.

WOOD ANEMONE or Wind Flower
(Anemone flos Adonis, Herba venita)

There are five sorts of Anemones, but I shall only describe that species which commonly grows in this kingdom. The leaves of the Wood Anemone are very much indented, the flowers are purple or reddish and the roots are hairy.

The roots when chewed – and they are a sovereign remedy for the expelling of phlegm – boiled in wine and laid upon the eyes clear the sight, and a decoction of it mixed with a small amount of barley drank by nurses increases their milk.

ANISE or Garden Anise
(Anisum)

This flowers and seeds in July, with the root dying every year after it has given seed. It grows frequently in gardens, with only the seed used in medicine.

It is an aid in expelling the wind from the body, and is exceedingly good to be given to infants in hot feeds, to prevent convulsions, the gripes and wind. The distilled oil is an excellent remedy against the pleurisy if it is applied outwardly. Anise can also be used to correct the effects of stronger purgative medicines, and the powder of it put into linen bags gives immediate help when applied to ruptures.

WILD ANISE
(Anisum agreste)

Wild Anise is commonly found in fertile, and sometimes in sandy soil. The leaves are somewhat rough and small, resembling the leaves of carrots and the seed is sweet scented. The virtues are the same as with the Garden Anise, but stronger and more powerful. The whole plant, but especially the seed, is a very great aid in promoting the discharge of urine.

APPLE TREE
(Malus saliva vel hortensis)

Apples comfort and cool the heat of the stomach, especially those that are somewhat sour. The leaves should be laid upon hot swellings, and they can also be applied to fresh wounds to prevent them turning bad.

ARBUTE or
Strawberry Tree
(Arbutus)

This is a small tree, not much bigger than a Quince tree, the body thereof is covered with a reddish bark, which is rough and scaly. The leaves are broad, thick and serrated, the flowers are white, small and grow in clusters, and after which follows fruit resembling strawberries, green at first, but yellowish afterwards, and at last red when ripe. It flowers in Spring and the fruit is ripe in Winter.

The fruit of this tree is of a cold nature, hurts the stomach and causes headaches.

WHITE ARCHANGEL
(Lamium album urtica mortua sive Arch-angellica flore albo non foetens folio oblongo)

This is the white-leaved, not Stinking Archangel. It grows everywhere by hedgesides, and flowers in April and May.

This herb can be recommended to be used for all swellings and it is also good against convulsions.

RED ARCHANGEL
or Red Dead Nettle or Stinking Archangel
(Lamium rubrum)

This grows in hedges near highways, and flowers in summer, and the whole plant has a strong, earthy and unsavoury smell. Both the leaves and the flowers can be used.

This is especially good to prevent excess menstrual bleeding, and indeed, is good against all inward bleedings. When outwardly applied, it is also servicible to cure wounds and inflammations, and when compounded with salt, it cures hard wens.

ARON, Arun, Wake
Robin or Great Cuckow Pint, spotted and unspotted
(Arum vulgare)

The leaves are long, large and of a shining green colour, shaped like the head of a spear. Sometimes they are full of black spots and on the stem grows a cluster of red berries, each containing one round seed. It grows in hedges and dry ditches flowering in May, and the berries are ripe in July.

The whole plant is very hot and biting, inflaming the mouth and the throat for a long time. The roots are pulverized and induce vomiting, and are good to clear obstructions of the lungs, and a poultice made of them and mixed with cow dung eases the pain of the gout.

ARSMART, sharp and hot, or Water Pepper
(Persicaria nomaculata urens vel hydropiper)

The leaves are long and narrow, much like the leaves of the peach tree, but not serrated about the edges, the flowers grow on short stems and cluster together. It grows in moist places near pools, or standing waters, and flowers in July and August.

The leaves have a hot burning taste, like pepper, and are good against cold swellings, upon being applied to them. The distilled water of this plant is exceedingly good against the stone, either in the kidneys or bladder. A decoction of it in water is good for old aches and pains. It is extraordinary good for the gout, or any arthritic disorder, for it raises a blister which carries off the malignity.

The dried leaves, made into a powder may be used with meat instead of pepper.

ARSMART, dead or spotted
(Persicaria maculata mitis)

The leaves are broader in the middle and larger than the former, they are smooth and have a dark brown or black semilunar spot in the middle of each of them, the flowers are of a carnation colour and the root is yellow and hairy. It grows in back yards and rich soil flowering about the same time with the former.

It cools fresh or green wounds and prevents inflammation, when the juice of the leaves is dropped into them. It is a cure for piles when roasted in the embers of a fire and applied with a little hot honey.

ARTICHOAK
(Cinara scolymus)

The decoction of the root drank, strengthens the stomach and confirms the place of natural conception in women, which, as it is reported, makes them apt to conceive male children, but care must be taken to remove the pith first.

Artichoaks are cleansers and are also good for the jaundice, and the leaves, when stamped and applied to the skin, draw thorns and splinters. A tea made of the leaves that grow on the stalk certainly cures the ague, when drank about an hour before the fit approaches, and taken three or four times.

COMMON ASARABACCA
(Alarum vulgare)

This has smooth, round leaves of a shining green colour, like ivy, but rounder and tenderer, while the flowers are of a brown purple colour, in the form of cups. It is planted in gardens and is chiefly used in Physick as an aromatic scent; they also have a sharp biting taste.

A tea made of it helps expel urine and is good for a cough, shortness of breath, convulsions, cramps, dropsy and sciatica. When mixed with honey it brings down the menstrual flux.

ASH TREE
(Fraxinus)

The leaves, bark and tender buds of the ash tree open up the liver, provoke urine and are useful against dropsy. The inward bark is given with success against fevers, and the wood, burnt into ashes, cures scabs and ringworms.

ASPARAGUS or
Sparagus, a corruption
of Sparrow-grass
(Asparagus sativus)

Eaten with oil and vinegar it provokes urine, and is good against the strangury, and a decoction of it opens obstructions of the liver and spleen, and is therefore good for jaundice and dropsy.

**MARSH
ASPARAGUS** or
Sperage
*(Asparagus sylvestris,
Corrundus)*

This herb is much like the former in its early stages, but the branches are rough and prickly when grown. It is found growing near the sea coast, and it has much the same virtues of the former.

ASP TREE or
Trembling Poplar
(Populus lybica tremula)

The leaves of this tree are almost round, they are deeply indented and are browner and harder than the leaves of the black poplar. They hang by long, but very slender stems, or root stalks, and this is the cause of their continual shaking. It grows in low, moist places.

The juice of the leaves assuages the pain of the ears, and heals ulcers of the frame, upon being dropped into them. A decoction of the bark provokes urine, relieves strangury and sciatica.

AVENS or Herb
Bennet
(Caryophillata)

The lower leaves are made up of about seven piunae which are hairy, as is the stalk, which grows to be about 2 ft high. The flowers grow on the tops of the branches on long-footed stalks consisting of five small yellow leaves, with several brown stamens in the middle, and the root is of a reddish colour which smells like cloves. It grows in woods and under hedges and flowers most part of the summer.

The roots when infused in wine render it more cordial and they ease pain arising from wind or cold in the bowels, and are also good against all kinds of haemorrhages or fluxes of the blood.

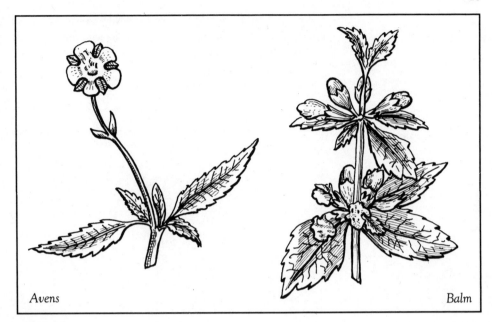

Avens　　　　　　　　　　　　　　　　　　　　　　*Balm*

B

BALM or Garden
Balm
(Mellissa)

Balm cures uterine disorders. When applied to wounds, the juice heals them. It invigorates the heart, the head and the nerves; it can be used with success against fainting, palpitations and failures of the heart, and is also effective against smallpox, measles and other malignant fevers. It can be applied externally to ease the stinging of bees and wasps.

BASTARD BALM
or Bastard leafed
Archangel
(Mellissae folio)

This herb has much the same virtues as the former.

BARBERRY BUSH
or Tree
(Oxyacantha)

The bark of this tree is of an ash colour, the branches are full of sharp thorns, the leaves are oval and indented and of a sourish taste, the flowers are yellow and are succeeded by cylindrical red berries. It flowers in April and May, and the berries are ripe in September. It grows in wild places and is commonly planted in gardens.

The inner bark which is of a deep yellow colour is good for yellow jaundice, and the fruit relieves diarrhoea.

Barley

BARLEY
(Hordeum polystichum)

This has a softening, cleansing and cooling nature; a decoction is very good for urinary fevers.

BASIL
(Basilcon vulgare majus et ocimum)

The stalks of Basil are four square and somewhat hairy, the leaves are of a yellowish-green colour, almost like the leaves of Mercury. On the tops of the stalks grow whorl-like spikes of white flowers, the leaves of which, when rubbed have a fragrant smell. It grows in gardens, flowering in July and August.

Basil invigorates the head and nerves, fortifies the brain and refreshes the spirits. It helps to cure old coughs and when bruised with vinegar and held to the nose, helps recovery from fainting. It is also good for inflammations of the eye.

SMALL BASIL
(Basilicon minus)

This is like Basil, but the leaves are much smaller and the stalks are round, bearing several little collateral branches on which grow leaves with jagged edges.

Its properties are similar to Basil.

WILD BASIL
(Basilicon sylvestre)

It has square hairy stems, leaves that are smaller than those of Small Basil, and the flowers are of a purple colour. It grows in sandy ground by river sides.

It arrests the bowels, the menstrual and other discharges.

COMMON GREAT BAY TREE
(Lauris vulgaris)

The juice of the beries when externally applied helps to cure deafness and hardness of hearing. An oil made from the berries is good for bruises. When taken internally the berries expel flatulence, comfort the head, mouth and nerves, and are also good against infections. A decoction of the bark of the root removes obstructions, promotes discharges of urine, breaks up bladder stone and is good for the kidneys.

BEAN or Great Garden Bean
(Faba major hortensis)

Beans, although having but little nourishment, have the special property of inducing diarrhoea and engendering flatulence. Bean meal applied outwardly dissolves tumours, while the water, distilled from the flowers, is cosmetic.

FIELD BEAN
(Faba equina sylvestris)

These are of no great medicinal value and are used to feed horses. However, they can be used externally as with the bean.

BEARS BREECH or Brank ursin
(Acanthus branca ursina)

The leaves of this plant are large, of a shining sad green colour and deeply cut. From among the leaves lying on the ground there arises a stalk about 2 ft high which bears leaves only near the top, which consists of a head of white gaping flowers standing among small and prickly leaves. It is cultivated in gardens, flowering in July and August.

The roots promote urine and ease cramps and pulmonary consumption. This herb can be applied to burns and dislocations. It is used in suppositories and also for stone and urinary crystals.

YELLOW LADIES BEDSTRAW
(Gallium luteum)

The leaves are small and narrow, set in a circle about the stalks, which are of a dark green colour, about 1 or 2 ft in height, on top of which grow thick spikes of small yellow flowers. It grows on banks and dry barren places, flowering in June and July.

When applied to burns, the crushed flowers alleviate the inflammation and when applied to wounds, they can heal them. Such a preparation is good to stop all kinds of haemorrhages including nosebleeds.

BEET
(Beta)

There are two sorts of Beet, white and red, which have the same virtues.

Beets cleanse the stomach. The juice, with a little honey sniffed into the nose, excites sneezing and clears the nose and brain, therefore easing a chronic headache. The juice also clears the ears if it is poured into them, and opens obstructions of the liver and spleen.

WHITE BEHEN or
Spratling Poppy
(Behen album polemonium)

It has tender stalks with large joints, the leaves are broad, are set opposite every joint and on the tops of the stalks grow white flowers which hang downwards. It grows frequently in cornfields and in meadows, flowering in summer.

The root is counted to be an antidote to poison, and is beneficial to the head and the heart.

White Behen

RED BEHEN or Sea Lavender Spike
(Limonium)

Because its leaves resemble the leaves of the lemon-tree, it is known in Latin as Limonium. The stalks are about 1 ft high, are bare of leaves, on which grow long spikes of small purplish-red flowers. It grows in salt-marshes, flowering in July and August.

The roots and seeds of Red Behen arrest the bowels and are useful against diarrhoea, dysentery or a large menstrual discharge. It assuages any pain in urination and is good for the spleen. The root, held in the mouth and chewed, takes away a toothache.

WATER BETONY
(Betonica aquatica scrophularia aquatica)

This herb is very like figworth, but the stalks are taller and the leaves larger in shape like Betony; the flowers are larger and of a redder colour, and it grows in watery places and near ditch sides.

The root is good for scrofulas tumours, piles, cancerous ulcers and itching. It also softens and disperses hard swellings.

WOOD BETONY
(Betonica sylvestris)

It is of a dark brown colour, bears blue flowers and is very strongly scented.

Infused in ale and drunk, it provokes periods and expels the afterbirth. It is very good against disorders of the uterus. It is also beneficial for barren women.

GARDEN BETONY
(Betonica hortensis vulgaris)

The leaves are dark green, serrated about the edges, the stalks are rough and square, about 1½ ft long and the flowers grow in spikes of a red–purple colour. It flowers in May and June.

It breaks down kidney stone, provokes urine and expels clammy and viscous phlegm from the lungs. If crushed and put up the nostrils it clears migraine headaches, and cures low spirits and giddiness. It removes obstructions to the liver, spleen and menstrual discharge. A dram of crushed leaves with honey and water is a powerful cure for cramps.

BILLBERRY BUSH
(Vaccinia)

The Billberry bush has long crenated leaves and black fruit. The fruit is known as whortes, whortle berries or billberries. It grows on mountains and bogs.

The berries are very cooling, good against burning fevers, inflammation of the liver and scurvy. They also arrest diarrhoea and vomiting. An agreeable syrup can be made from the juice of the berries for the same remedies.

GREAT BINDWEED
(Convolvulus major albus vel volubilis major)

This herb has tender stalks and branches, winding and climbing about anything in its way; upon these grow soft leaves, almost like the leaves of ivy, but much smaller, and on the tops of the branches grow white bell-fashioned flowers. It grows in hedges, flowering in June and July.

This plant is of no great use in medicine, although the root can be employed are a purgative.

BIRCH TREE
(Betula)

The liquid that is drained off this tree in the springtime is good for dispelling urinary disorders, like stones, pains and bleeding. A decoction of the leaves, when drunk, is considered good for scurvy.

BIRD LIME or ordinary Mistletoe
(Viscus quercinus vel viscum)

This takes root in the branches of trees; the berries which grow on it are round, white and pellucid.

The leaves and berries cure swellings, abscesses and sores. The seed, pounded with white wine lees, and applied to the side, mollifies the hardness of the spleen. It is very good for the head and nerves, for apoplexy, convulsions, paralysis and giddiness. For this purpose it can be hung around the neck of the patient, or drunk in black cherry water.

LONG BIRTHWORT
(Aristolochia longa)

It has several square slender branches about 9 in long, about which grow leaves like the leaves of ivy; the flowers are pale purple with a strong scent, and these are succeeded by roundish fruit, as big as walnuts. The root is almost as thick as a man's wrist, about 1 ft long, of a yellowish colour and a bitter taste. It is planted sometimes in gardens, but commonly grows in woods.

It provokes urine and menstruation and drunk with pepper and myrrh, it expels the dead child, the afterbirth and all superfluities of the womb. There is nothing better for curing rickets in children, either used externally or internally.

ROUND BIRTHWORT
(Aristolochia rotunda)

The stalks and leaves are like the former, except that the leaves of this are somewhat rounder, the flowers are longer and narrower, and of a faint yellowish colour. The roots are round like a turnip. It is planted in gardens and grows wild in ditches and several other places.

The root is the only part used and is of value against convulsions, such as cramps and epilepsy. It clears the bowel and phlegm pretty smartly. It provokes menstrual discharge, assists birth and expels the afterbirth. It scours corrupt and dirty ulcers; it draws out splinters and broken bones. A hot, wet application made from it helps old aches, bruises and dislocations.

BISHOP'S WEED
(Ammi vulgare)

The leaves of this herb are large and indented, divided into several other long narrow leaves; the stalks are green and round, on the top of which grow large clusters of small white flowers. The seed is like parsley seed and has a hot taste. It grows in gardens, flowering in July and August.

The seed is very good against colic pains; it provokes urine and menstruation as well as flatulence. Mixed with honey, it disperses congealed blood.

BISTORT the
Greater, or Snakeweed
(Polygonum bistorta)

The leaves very much resemble the leaves of common dock, but are longer and more slender; the stalk is long, smooth and tender, and the flowers grow in spikes, like ears of corn, and are of a carnation colour. It grows in moist meadows, flowering in May.

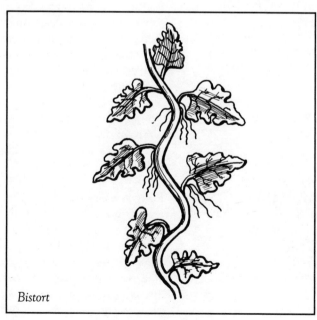

Bistort

A decoction of the root is very good against diarrhoea, dysentry and haemorrhages. If a powder of it is taken with red wine, it stops vomiting. A decoction of the leaves is very good against all sores, and inflammation of the mouth and throat. It also secures loose teeth.

BLITES or Goosefoot or Sowbane
(Plitum album majus)

It grows about 2 ft high and has thick stalks, clothed in many leaves, like beet leaves. After the flowers come the seed remains enclosed in little flat husks. It is planted in gardens and flowers in July.

Only the leaves are used in medicine; they are cooling and softening and are sometimes put into suppositories.

SMALL or COMMON BLUE BOTTLE
(Cyanus minor vulgaris)

It has many slender whitish angular stalks, on which grow narrow sharp-pointed and indented leaves; on the top of the stalks grow small scaly buds, from which come pleasant blue flowers. It grows amongst the corn, flowering in June and July.

The distilled water of the flowers is excellent for sore, bloodshot and inflamed eyes, while an infusion of them is beneficial for jaundice. It is good for all sorts of obstructions.

BORAGE
(Borrago)

It grows in gardens and flowers in June.

It comforts the heart and makes people cheerful and merry, therefore driving away sadness and melancholy. It is an antidote to poison and good against malignant fevers, smallpox and measles. Boiled with honey and water, it is good for hoarseness.

BOX TREE
(Bauxus)

The leaves of the Box are hot, dry and astringent.

An oil distilled from the wood is used for toothache, a little lint or cotton being dipped into it and put in the hollow tooth.

BRAMBLE
(Rubus vulgaris major fructu nigra)

It grows everywhere in hedges. The undertops and leaves are astringent, as is the unripe fruit.

The tops and young buds cures sores and ulcers of the mouth, throat and uvula, and held in the mouth and chewed, they secure the teeth. A decoction of them is effective in stopping diarrhoea, menstrual discharge and any flow of blood. The roots provoke urine and break up bladder stone, and the crushed leaves cure piles.

Bramble

White Briony

WHITE BRIONY
(Bryonia alba)

The leaves are something like vine leaves, but more rough and hairy. It has a great many rough, tender branches climbing on the hedges in its way. The flowers are of a whitish–green colour and are succeeded by round berries, which are in the beginning green, but afterwards red. It grows in lanes and by hedgesides, and flowers in May.

The root of White Briony powerfully purges viscous phlegm, and watery humours, both upwards and downwards. It is also excellent for all nervous disorders. An oil made from the root cures freckles and spreading sores, and if it is pounded and mixed with wine, it draws out splinters and broken bones.

BLACK BRIONY
(Bryonia nigra)

The leaves are almost like the leaves of bindweed, the stalks and branches twine and climb about hedges and trees, and the fruit clusters together like small grapes. It grows in the same places as white bryony and flowers about the same time.

The root is good for epilepsy, it provokes urination and menstruation and it cleanses the kidneys. The tender sprouts or tops of it, eaten in salads, will cleanse the stomach. It is very dangerous for women with child to make use of any part of it.

BROOKLIME
(Anagallis aquatica, Becabunga)

This herb is so well known that I need not describe it. It flowers in June, holding the leaves all winter, and grows in moist ditches and near running waters.

It is good for breaking up stone and gravel, and for easing the passage of urine. Applied as a poultice, it dissolves tumours or hard swellings of the liver and spleen.

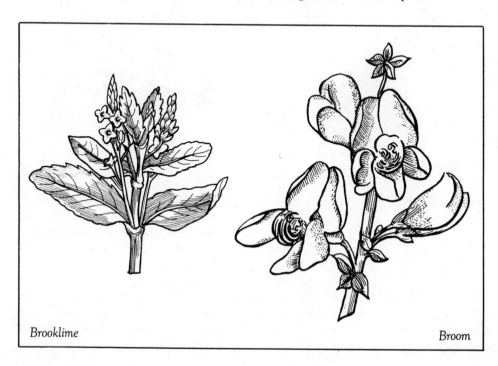

Brooklime

Broom

Butcher's Broom

BROOM
(Genista)

This flowers in April and May.

The tender branches, leaves and buds of Broom, boiled in wine or water are very good in removing any obstruction of the liver, spleen, kidneys or bladder. The seed has the same property, a dram or a dram and a half to be taken at a time. The flowers, if mixed with Hog's lard, assuage the pain of the gout. The ashes infused in white wine powerfully provokes urine, and are good for jaundice.

BROOM-RAPE
*(Orobanche,
Rapum genistae)*

It grows about the roots of Broom.

A decoction of it in white wine is excellent for breaking up stone and expelling gravel and provoking urination. The fresh juice or oil of broom-rape heals all corrupt and rotten ulcers.

BUTCHER'S BROOM
*(Bruscus ruscus sive
oximysiner)*

This grows in gardens and flowers in summer.

It is mainly the root that is used in healing. A decoction of it breaks up stone, expels gravels and aids urination. It is useful against jaundice and provokes menstrual flow.

BUCKBEAN
(Trifolium paludosum)

Buckbean is so well known that it needs no description. It grows in marshy and soggy ground, flowering in May and June.

The seed is effective against a cold, coughing, diseases of the lungs and spitting of blood. Taken with honey and water, the leaves are good against scurvy, gout and rheumatism.

BUCKTHORN-TREE
(Rhamnus catharticus, Spina cervina)

This grows in woods and hedges, flowering in June, with the berries ripening about the end of September.

Made into a syrup, the berries gently purge phlegm, and are good against scurvy and gout. Ashes from the wood made into lees, cure scabs and itches.

BUCKTHORN
Plantine
(Coronopus vulgaris, Cornu cervinum)

The long and narrow leaves lie flat on the ground in a circle — from whence it is called Stella terrae, or Star of the Earth. The spikes are narrow and consist of four leaved flowers, growing on hairy stalks 3 or 4 in long. It grows in sandy ground, flowering in June.

It has a drying and binding quality when applied to wounds. It is extraordinarily good against the bites of venomous creatures, especially the bite of a mad dog.

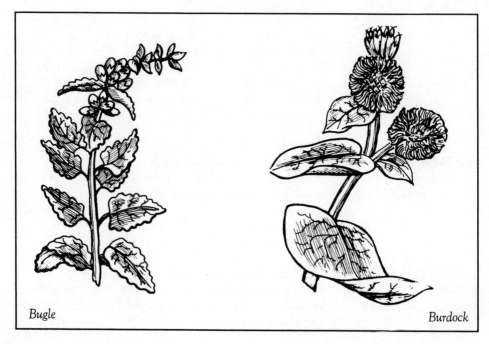

Bugle

Burdock

BUGLE or Middle
Confound
*(Bugula,
Consolida medi)*

Bugle creeps on the ground like moonworth. It has long broad leaves on square stalks with few leaves positioned opposite to one another, which are of a dull green colour. The blue flowers grow in loose spikes, whorl-fashion. It grows in woods and hedges, and flowers in May.

It dissolves clotted or congealed blood, opens obstructions of the liver and gall, and cures ulcers and sores of the mouth. If crushed and applied to wounds, it cures them.

**GARDEN
BUGLOSS**
*(Buglossum Hortense vel
vulgare)*

When fully grown the stalks reach 2 or 3 ft in height. The stalks bear long narrow leaves, and on top grow rough thistly heads bearing purplish flowers. These flowers are succeeded by four-cornered rough seed. It flowers in June and July.

This is a great cordial herb, and it has the same properties as Borage; it is a herb which uplifts the heart and spirits.

WILD BUGLOSS
(Buglossum sylvestre)

This is a much smaller plant than the former, not growing above 1 ft high. The stalks are thick, succulent and prickly, the leaves long and narrow, and the flowers and seed are like those of the Garden Bugloss. It grows near hedges, highways and among corn, flowering in May.

It is seldom made use of in medicine, although it is considered to have the same properties as Garden Bugloss.

BULLACE-TREE
(Prunus sylvestris major)

Wild plums are astringent, of a dry cold nature, their juice stops diarrhoea and flows of blood, and may be used instead of Acacia.

**GREAT COMMON
BURDOCK** or
Clot Bur
*(Bardana major vel lappa
major)*

The plant grows in rich soil, such as backyards, flowering in June and July.

The leaves pounded with salt are very effective when applied to the bite of a mad dog, as is the juice of the leaves drunk in wine. A dram of the root pounded with kernels of pineapple is a powerful remedy for the spitting of blood. A small amount of the seed pounded and drunk with white wine, eases bowel disorders. It promotes urine and breaks up bladder stone. The leaves boiled in milk and applied as a poultice are good for gout. The roots cause perspiration and are good against poisons and fevers. The green leaves pounded with the white of eggs cure old sores and burns.

BURNET
(Pimpinella minor hortensis, Sanquisorba)

The stalks grow to about 1 ft high; they are downy and of a reddish brown colour. The leaves are feather-like; on the tops of the stalks grow round firm heads, consisting of a cluster of small reddish four-leaved flowers.

A decoction of this herb is very good against dysentry, spitting or pissing blood, menstrual disorders and all flows of blood. If the green leaves are crushed and applied to wounds they will prevent abscess and inflammation. If steeped in wine they cheer the heart and exhilarate the spirits, for, in short, it is an excellent cordial herb.

WILD BURNET
(Pimpinella vel savifraga major)

It grows in meadows and in old stone walls.

It is excellent for provoking urine and for expelling stone and gravel and flatulence out of the stomach.

BURR-REED or
Burrflag or Reed Grass
(Sparganum ramosum)

It has long narrow leaves, sharp on both sides, with a sharp crest raised up so that they almost seem triangular. The stalks are 2 or 3 ft high, upon which grow round prickly heads, or burrs, as big as nuts.

It is of a cold and dry nature. A decoction of the rough burrs of this herb in wine is good against bites of venomous creatures, either if it is drunk, or if the wound is washed therewith.

BUTTERBUR
(Pelatites vulgaris)

It grows by riversides and by marshy ground, the flowers appearing in early March.

The roots of the herb powerfully provoke sweat and are therefore good against all malignant fevers. They also provoke urine and kill worms in children. Pulverized and applied to wounds, inflammations and ulcers, they act as a cure.

BUTTER-WORTH
or Mountain Sanicle
(Pinguicula gesneri seu sanicula montana)

This grows in boggy moist meadows and has the same virtues as Common Sanicle.

C

CABBAGE and Coleworts
(Brassica sativa, Caulis)

All coleworts have much the same properties, in that they make the belly soluble.

The juice with ground fenugreek helps gout, and heals old rotten sores. Mixed with vinegar, the juice put warm into the ears is good against deafness, and for the ringing humming noise in them. The stalks, eaten raw, prevent drunkenness; the young leaves are cooling and good for laying on inflammations. The ashes of the stalks mixed with hogs lard, are good for applying to old pains and aches in the side. The leaves are great drawers and purges.

COMMON MOUNTAIN CALAMINT
(Calamentha vulgaris montana)

It is found growing on mountains near hedges and highways, flowering in June and July.

It is very good for all internal bruises, clotted and congealed blood, shortness of breath, and for obstructions of the lungs and the jaundice. It provokes urination and menstruation and expels wind from the stomach. It can also be used to expel a dead child from the womb. A decoction of it with honey and salt expels all kinds of worms out of the body.

COMMON CALAMINT of the Shops.
(Calamentha officinalis)

This Calamint is called Wild Pennyroyal. The stalks are square and downy, on which grow triangular leaves, opposite one another. The flowers grow among the leaves; they are of a pale purple colour and there are sometimes three or four on a stem. It grows near hedges and highways and has much the same virtues as mountain calamint.

WATER CALAMINT
(Calamentha aquatica)

It grows about 1 ft or more high, the stalks are square and hairy and the leaves are larger and longer than the Common Calamint. Indented about the edges, the flowers grow in very thick whorls, with the leaves on the upper part of the stalks. It grows in moist places where water has stagnated in winter and it flowers in June. It has the same qualities as mountain calamint.

CALTHROPS or the ordinary Star-Thistle
(*Calcitrapa,
Carduus stellatus vulgaris
foliis papaveris erratici*)

The lower leaves grow flat on the ground encompassing the root in a circle; the stalk is about 2 ft high and is divided into numerous branches, and the flowers are reddish or purple. It grows near highways and on commons, flowering in June.

The root is a powerful remedy against stone, gravel cholic and convulsions. An ointment made together with foxglove, hemlock and dropworth cures abscesses and sores.

CALVES' SNOUT
or Snapdragon
(*Antirrhinon sylvestre*)

This herb has straight round stems full of branches and leaves that are not unlike the leaves of Pimpernel. The flowers are like the flower of Toad Flax, but are much larger and are of a faint yellow colour and are succeeded by long husks, the front part of which resembles a calves snout. It grows in gardens, flowering in July and August, and is also found growing wild. There is a lesser Calves Snout (Orontium), which grows wild in fields, near highways and under hedges.

It is thought that whoever carried about a Calve's Snout cannot be hurt by any poison or poisonous creature. The lesser Calve's Snout is hot and dry and a decoction thereof washes away the yellow colour which remains in the body after jaundice.

CAMOMILE
(*Chamaemelum*)

This flowers in June and July.

It provokes urination and menstruation, expels a still-born child, aids flatulence and eases bowel disorders. A decoction of it applied to the area of the bladder and kidneys expels gravel and eases pain. An oil made of it is good for stitches, pains and old aches of the limbs, and dislocations. The flowers help to expel wind, and a powder made from them and mixed with wine is a good cure for an acute fever. Boiled with vervain in milk, and applied to an old pain or ache, camomile will help cure it. An ointment made of it with fresh butter and goosedung is an excellent remedy for a burn or a scald.

DOG'S CAMOMILE
or Stinking May Weed
or Wild Camomile

This has a thick green stem, full of juice, which quickly breaks underfoot; the flowers consist of broad white petals set about a yellow fistular centre. It grows frequently in

*(Cotula faetida,
Chamaemelum faetidum,
Cynobotane)*

waste places and among corn, flowering in May and June.

It is good for women with the falling down of the womb, if they but wash their feet with a decoction of it. A hot wet application of it will ease the swelling of haemorrhoids.

WILD CAMPION
(Lychnis sylvestris flore rubro)

It has rough white stems, the leaves are downy, oval and sharp pointed. The flowers grow on the tops of the branches and consist of five red round sharp-pointed leaves. It grows in ditches and moist places, flowering in May. There is another sort of wild Campion which bears white flowers (Lychnis sylvestris flore albo) which in all other respects is like the former.

Both types are very good against the bites of venomous creatures. A small quantity of seed will eliminate bile. The red flowers are good against an excessive menstrual flow.

CARAWAYS
(Carum sive careum)

It is very good for helping digestion, expelling all kinds of wind and flatulences, and is therefore good for bowel disorders. It is useful for headache and weakness of sight. It is like annise seed in operations and virtues.

Camomile

CARLINE THISTLE
*(Carlina,
Chamaeleon albus humilis)*

It bears long rough narrow and prickly leaves, deeply cut on both sides. The stem is not more than a hand and a half high. The flowers consist of a border of white shining, sharp-pointed petals set about a yellow fistular centre. It flowers in July and August.

The root provokes urination and menstruation and is an antidote to poison. If a powder made from it and the quantity of a dram is taken inwardly, it is an excellent protection against disease, for, as we may read, the army of the Emperor Charlemaigne was, by the help of this root, preserved from pestilence. Held in the mouth, it is good against toothache. Being externally applied with vinegar, it cures the scurf and itch.

CARROT
(Pastinaca tenuifolia sativa)

Carrot roots when eaten are indifferent nourishment, and usually provoke urination and excite sexual desire. If made into a powder and drunk with mead, they open obstructions of the liver, spleen and kidneys, and are therefore good against jaundice, and gravel.

WILD CARROT or
Bird's Nest
(Daucus vulgaris, Pastinaca silvestris tenuifolia)

The wild carrot is not unlike the garden carrot, except that the leaves are more finely divided, rougher and more hairy. The stalks grow 2 or 3 ft high, upon the tops of which grow small white flowers and when these fall off they close themselves into a hollow round form like a bird's nest. It grows frequently in pasture grounds and fallow fields, and it flowers in June.

Wild Carrot

The seed provokes urination, menstruation and the breaking up of stone. It is believed that it makes women fruitful if taken often. A handful of finely chopped leaves boiled in 2 or 3 pints of milk whey will expel the worst attack of gravel and ease urination. If the flowers are not available, you may use the root.

CATMINT or Nep
(Nepeta,
Mentha selina)

This has square hairy soft stalks, full of joints; the leaves are like those of Dead Nettle, and the flowers, which are white, grow on the tops of the stalks in long whorled spikes. It grows in gardens and sometimes wild, in lanes and hedges, flowering for most of the summer. It is called Catmint because cats are very fond of it, especially when it is withered, for then they roll themselves in it, and chew it with great pleasure.

It provokes urination and menstruation; it expels the still-born child; it opens obstructions of the lungs and the womb, and it is good for internal bruises and shortness of breath. Drunk with salt and honey, it expels worms from the body.

GREAT CELANDINE
(Chelidonium majus,
Hirundinaria)

It grows in gardens and sometimes wild among rubbish, on walls and old buildings, and it flowers in May.

It is an excellent medicine to preserve the eyesight; it removes inflammations, specks and films from the eyes, and

Catmint

Great Celandine

it can also strengthen them. The root removes obstructions of the liver and spleen, and cures jaundice. When it is chewed, it eases toothache. A strong decoction is very good against skin eruptions, ringworms and scorbutic sores.

It is called Chelidonium — in Greek, Swallow Herb — because, as Pliny wrote, it was first discovered by swallows when it restored sight to their young ones.

SMALL CELANDINE or
Pileworth
(*Chelidonium minus*)

This is a low herb with small brownish stems; the leaves are small and somewhat round; the flowers are yellow and the root has several little whitish tubers. It grows in meadows and moist pastures, and flowers in April.

It provokes vomiting and a discharge of phlegm from the chest. If a decoction of it is gargled, or if the juice of the root mixed with honey is sniffed into the nostrils, the brain is purged and a stoppage of the nose unblocked. The root, pounded in a little wine or urine and applied to piles, dissolves and heals them.

Small Celandine

LESSOR CENTORY
or Red Ordinary Small
Centory
(*Centaurium minus*
vulgare flore purpureo)

A decoction of it purges bilious and phlegmatic conditions. It removes obstructions of the liver and spleen and it eases cramps and convulsions. The juice dropped into the eyes with honey clears them. It is also good for the stomach.

It may also be used as part of an infallible cure for the cholic. Take this:

Two handfuls of century, agrimony and camomile flowers. One ounce each of gentian roots and carduus seed. Two drams each of cubebs, galingal roots, Cardamum, nutmegs, cinnamon, cloves, mace. Half an ounce of caraways seed, civil orange peel, lemon peel. One handful of elder berries, and a few grains of whole pepper, all infused in a gallon of aqua vitae.

YELLOW CENTORY
(*Centaurium luteum*
perfoliatum)

The root conglutinates and heals all fresh wounds, and it stops all kinds of flows and haemorrhages.

It takes its name from the Centaur Chiron who, by using it, cured himself of a wound he received from one of the arrows of Hercules.

CHARLOCK or Wild
Mustard
(*Rapistrum*)

It flowers from April to Midsummer.

This herb, but especially the seed, is hot and dry. It is boiled and eaten by the common people in Spring, instead of Coleworts. Made into a poultice, it is good against cramps and convulsions.

RED CHERRY-TREE
(*Cerafus vulgaris rubra*)

This tree bears red cherries (Cerafa rubra), which are of a cooling moist nature. They purge and comfort the stomach, assuage thirst, and ease the conditions of stone, gravel and epilepsy.

BLACK CHERRY-TREE
(*Cerafus nigra*)

These cherries are good for all uneasiness of the head and nerves, such as epilepsy, convulsions and paralyses. They also provoke urination and break up stone, and in general, the distilled waters of these cherries is of great use in medicine.

COMMON GARDEN CHERVIL
(*Chaerefolium vulgare*
sativum)

It is a small low plant, with winged leaves, smaller and finer than Parsley, and on the tops of the stalks grow small clusters of five-leaved white flowers. Its virtues are much the same as those of Parsley.

A handful of chervil squeezed into ale or whey and drunk

is very good for pleurisy. It provokes urination and breaks up stone. It cures rickets, and helps rheumatic pains, used externally and internally. It is also useful for stoppping vomiting and diarrhoea.

CHESTNUT TREE
(Castanea)

It is thick-set with long, narrow and sharp-pointed leaves, deeply serrated around the edges, while the catkins are long, thin and slender. It is frequently planted in gardens and parks. Chestnuts, although quite nourishing, are hard to digest.

They are good for chest problems, and for arresting the bowels. A paste made from them is good against coughing and spitting of blood.

EARTH CHESTNUT or
Earth-Nut or Pignut
(Bulbocastanum)

This plant has a root as big as a large nutmeg; the leaves are finer and smaller than those of meadow Saxifrage and on the tops of the branches grow thin clusters of small white flowers. It grows in sandy places, flowering in May.

The root, which is the only part used, when roasted has a pleasant taste and is nourishing. It promotes sexual desire and also eases dysentery and menstrual flow.

CHICKWEED
(Alsine vulgaris)

Boiled in water and salt, it is a powerful remedy against the heat of scurvy, and the itching of hands. It is cooling and moistening, good against inflammations, boils, hot

Earth Chestnut   Chickweed

Cinquefoil

swellings and pains in any part of the body. For these remedies, the juice, or a poultice with hog's lard is applied.

CINQUEFOIL
(Pentaphyllum vel quinquefolium)

The stalks lie on the ground, emitting small fibrous roots from the joints, by which it easily propogates itself. At every joint grow five leaves, set together on a 1 ft stalk. The flowers are yellow and consist of five leaves. It grows in hedges and by waysides, flowering all summer.

The roots of Cinquefoil boiled in water until a third part of the water has been boiled away, ease the raging pain of toothache, if held in the mouth. They help to cure dysentery and all other diarrhoeas. Fevers can be cured by giving a small amount of root-powder two or three times a day. The juice of the leaves mixed with honey and water, or in vinegar whey has the same effect.

GARDEN CLARIE
(Horminium hortense)

It flowers in June and July.

It is of a hot and dry nature. The seed, pounded and tempered with water, draws out thorns and splinters, and dissolves all kinds of swellings. The leaves infused in wine, comfort a cold windy stomach. Fried with eggs they strengthen the kidneys and the back.

WILD CLARIE
(Horminium sylvestre ceu oculus Christi)

Wild Clarie bears wrinkled and serrated leaves; the stalks are square and somewhat hairy while the flowers are of a deep blue colour, like the flowers of Lavender. It grows in gravelly ground, flowering in June and July.

The seed mixed with honey cures inflammations of the eyes and cleanses and strengthens them. If drank in wine, the seed excites bodily lust.

CLIVERS or Goose Grass Clivers
(Aparine)

It grows in hedges.

It provokes urine, breaks up stone and gravel, and sweetens the blood. Being externally applied, it stops the bleeding of fresh wounds, and it is also good against scrofula, and lumps on the skin. It eases stitches, and disperses cold blockages in the joints. It cures the pain of the ears if the juice is dropped into them.

It is known as Goose Grass Clivers, because it is exceeding good to fatten geese.

CLOVE JULY Flower
(Caryophyllum rubrum), also Double Clove July Flower *(Caryophyllum multiplex),* and The Great Garden July Flower *(Caryophyllum altile majus)*

These flowers have a pleasant aromatic smell and are cultivated in gardens, flowering in July.

Made into a syrup, they are useful for all diseases of the head and nerves. They are also a good antidote to poison, as well as being beneficial for the heart. Toothache will be cured with a medicine which is made up from ½ oz of refined oil of Clove July, a little camphire and ½ oz of four times-distilled spirits of Turpentine. This liquor will infallibly cure the toothache, if you dip a little cotton into it and apply it to the aching tooth.

CLOVER-GRASS or the Great Trefoil
(Trifolium majus purpureum sativum)

It is of a hot and dry nature. The leaves and flowers, or the seed infused or boiled in water and taken internally, are good against pains of the side, and uterine disorders, as well as provoking urination and menstruation. It is a good antidote against acute fevers and poisons as well as hysteric fits.

CLUBMOSS or Wolf's-Claw
(Museus clavatus, Lycopodium)

It spreads on the ground, sending out several distinct branches, which are thickly set and covered with a kind of hair which is of a changeable colour, between green or yellow. These branches send out other branches, the ends of which resemble the claws of a wolf. It grows in unmanured fields and woods.

If inserted into the nose, it stops bleeding, and is a useful ingredient of astringent ointment. A decoction of it in claret helps stop diarrhoea.

COCKLE or Bastard Nigella
(Lychnis segetum, Melanthium sylvestre)

It has a striped stalk of about two spans long; the leaves are ashen-coloured resembling the leaves of dill. The flowers are like the flowers of garden Nigella, only bluer, and it grows in fields, flowering in June and July.

The seed helps to provoke menstruation, dissolve flatulence, expels worms from the body and cure shortness of breath. A decoction of it in water and vinegar held in the mouth, eases toothache.

WILD COLEWORT
(Brassica sylvestris)

It is of a hotter and drier nature than the large Coleworts, stronger and more purgative and cleansing, and is therefore not to be eaten. When pounded, the leaves heal wounds, dissolve tumours and draw out rottenness from running sores and from bilious conditions. If made into a syrup with sugar and liquorice, it cures an abscess of the lungs.

Colts-Foot

COLTS-FOOT or Folefoot
(Tussilago farfara)

It grows in moist places, and flowers at the end of February.

It is good for all diseases of the lungs, for coughs and consumption. The leaves, pounded with honey, cure all skin inflammations.

GARDEN COLUMBINE
(*Aquilegia*)

It is good for obstructions of the liver and spleen, for jaundice, sore mouths, and inflammations of the jaws and throat.

WILD COLUMBINE
(*Aquilegia sylvestris*)

It grows like Garden Columbine, but the flowers are of a paler blue and the leaves are longer. It grows in woods and flowers in May and June.

It cures ulcers, and itchy and scabby skin, but especially the mange on cattle or dogs.

COMFREY
(*Symphitum,
Consolida major*)

It grows in gardens, but sometimes in ditches and watery places, and it flowers in June and July.

The roots are exceeding good for all them that spit blood, for they heal all inward wounds and ruptures, and also to clear up and purge phlegm. They also glutinate and heal external wounds, and are good for hot tumours and inflammations. The roots, beaten into a poultice, ease gout.

CORN-SALLET or
Lamb's Lettuce
(*Lactuca agnina*)

It grows on the borders of fields, by waysides and in cornfields, flowering in July and August.

It provokes urine, promotes sleep and alleviates pain. The juice, mixed with breast-milk, cures burns and clears the sight.

COSTMARY or
Alecoath
(*Balsamita mas,
Costus hortorum,
Herba divae mariae*)

It has hard round slender stalks, about 1 ft high; the leaves are light green, serrated about the edges and the flowers are like the flowers of Tansy, but smaller. It has a pleasant strong smell and is of a bitter taste. It is planted in gardens, and flowers in July.

It heals colic pains, stops vomiting and the looseness, and the spitting and pissing of blood. It strengthens the stomach, expels wind, opens obstructions of the liver, spleen and brain, and stops superfluous catarrhs.

COUCH-GRASS or
Dog's Grass
(*Gramen cranium*)

It provokes urine and opens obstructions, especially of the kidneys and bladder. A decoction of it is useful for colic pains and painful urination. The leaves and roots, if pounded, heal wounds and prevent bleeding.

COWSLIP or Paigle
(*Paralysis,
Primula veris major*)

Cowslips grow in gardens and moist meadows, flowering in April.

They purge the stomach, promote sleep and are good

against epilepsy, paralysis, apoplexy and headaches. Tea is commonly made from the flowers to promote sleep and a salve made from the leaves strengthens the nerves.

CRAB-TREE or Wilding
(Malus sylvestris)

Crab-tree grows frequently in hedges, and flowers in April and May, with the fruit ripening in September.

The juice of crabs is a useful ingredient in astringent gargles for ulcers of the mouth, throat and also for dropped uvulas. It is also used for burns, scalds and inflammations.

CRANES-BILL or Doves-Foot
(Geranium colum binum pes columbinus)

It has tender reddish hairy and slender stalks, the leaves are small and round, while the flowers are small, and of a pleasant pale red turning to purple, which are succeeded by long heads, each resembling the head and bill of a crane. It grows among weeds, in rocky places and by hedgesides, flowering in May and June.

It provokes urine and dissolves gravel and stone. It is also very good in stopping all kinds of flow.

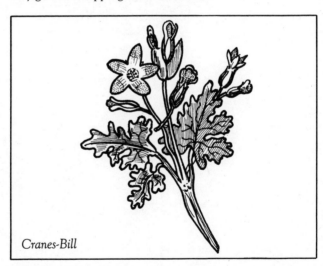

Cranes-Bill

GARDEN-CRESS or Tongue Grass
(Nasturtium hortense)

It grows in gardens, and flowers in May.

The leaves are good for the nerves. and for curing diseases of the mouth. The seed purges the stomach, expels worms, opens obstructions of the spleen, provokes menstruation and expels a dead child and the afterbirth. Made into a poultice with honey, it cures hardness of the spleen, swellings and scabby skin.

WATER CRESS or
Watergrass
(Nasturtium aquaticum)

It grows in moist ditches, standing waters and springs, and flowers in June.

It provokes urine, breaks up stone and gravel, and is useful against jaundice.

SWINES CRESS
(Coronopus ruellii)

It grows near highways and on banks, flowering in May and June.

Its nature is hot and dry, being astringent. The root, roasted in embers and eaten is good against a flux proceeding from the coldness of the stomach, and it is also useful in dispersing scrophulous tumours.

WILD CRESS or
Churles Cress
(Nasturtium sylvestre vel rusticum)

It bears long large leaves, cut about the edges; the stalks are round, about 1 ft long and divided into several branches, on which grow small husks like those of Shepherds Purse, and which contains sharp biting seeds. It flowers at the end of Spring and is in seed all Summer.

It is of a violent hot and dry nature, especially the seed. It purges violently, both upwards and downwards, and also provokes menstruation and breaks up internal abscesses. If applied externally, like mustard seed, it is good for sciatica.

Cuckow Flowers

Crow-Foot

CROSS WORT
(Cruciata)

It has many square stalks, full of joints, the leaves are small, broad and round-pointed, with four leaves growing at every joint, opposite one another in the form of a cross; the flowers are yellowish, and consist of four leaves. It grows near trenches, water courses and in hedges, and it flowers in July.

It heals and conglutinates wounds. It is also good for ruptures, but the herb must be boiled before it is applied to the grieved part.

CROW-FOOT or
Common Creeping
Crow Foot or
Buttercup
*(Ranunculus pratensis
repens)*

There is also round or knobbed rotted Crow Foot, Water Crow Foot and Round-leaved Water Crows Foot. All these species of Crow foot have much the same properties, being caustic. If pounded and laid on any part of the body, they raise blisters. If the roots are pounded and sniffed into the nose, they provoke sneezing. The pounded leaves can be applied beneficially to eruptions, scabs and warts.

**CUCKOW
FLOWERS**
*(Cardamine,
Nasturtium pratense
majus)*

Its leaves are pinnated, the stalks smooth and round, and the flowers consist of four roundish white leaves. It grows in meadows and flowers in April.

It provokes urination, breaks up stone and is good for the kidneys and nerves. It is also beneficial against scurvy, cancerous ulcers, freckles, epilepsy and the pissing of blood.

GARDEN CUCUMBER
(Cucumis sativus)

This is cold and moist in nature, and it is therefore good for inflammations of the stomach and bowels. It yields bad and small nourishment, and it is not good to eat too much of them, for they fill the veins with fevers and other diseases. The green leaves crushed with wine heal dog-bites. The seed is cooling, it provokes urine and breaks up stone.

An infallible cure for pleurisy consists of taking unpeeled cucumbers, slicing them into a frying pan, covering them with salad oil, and cooked on the fire until they are well-fried. Do this thrice, but retain the cooked oil. This oil can also be rubbed onto stitches in the side.

WILD CUCUMBER or
Squirting Cucumber
(Cucumis agrestis, Afininus)

The leaves are rougher and smaller than the leaves of Garden Cucumber; the stalks are round and rough, creeping on the ground, while the flowers are pale yellow with each flower consisting of one single leaf. The fruit is as big as a large olive, which if gently pressed, will squirt out with great violence. It grows in gardens and wild places, and it flowers in July.

The dried juice is a great purgative, and if mixed with oil and outwardly applied, will heal inflammation of the throat. The juice of the leaves can ease the pain of the ears and the root steeped in vinegar alleviates the pain of gout.

CUDWEED or Herb Impious
(Gnaphalium vulgare, Filago, Herba impia)

The leaves are small and covered with a fine woolly substance; it has one woolly stalk about 1 ft high, and it bears naked yellow flowers growing together in clusters. It grows in dry barren places near the sea coast, and in fallow fields.

It has a dry astringent nature. A decoction in red wine is good against dysentery, haemorrhages and all kinds of flow.

CUMMIN
(Cuminum vel cyminum)

It bears fine slender leaves like Fennel, only much smaller; the flowers are reddish white and grow in small clusters, and it is about 1 ft high, while the seed is long and brown and it is sown yearly in gardens.

Cummin seed is useful for expelling wind from the stomach and bowels, and for clearing obstructions of the lungs, it having a hot and dry nature.

CYPRESS TREE
(Cupressus)

It is planted in gardens for its pleasant verdure, being a perennial or evergreen.

It stops bleeding and heals wounds. A decoction of the cones, taken internally, is very good against dysentery, diarrhoea, haemorrhages and the spitting of blood, and also vomiting and involuntary urination.

Cudweed

D

GREAT DAISY
(Bellis major)

It bears long leaves, serrated about the edges. The stalks are about 1 ft high and the flowers are composed of several broad white petals set about a broad yellow centre. It grows in meadows and moist pastures, and it flowers in June.

It is of a cold moist nature, and is good against burning ulcers, abscesses, inflammations and wounds. It is excellent in curing diseases of the lungs, and also for coughs, shortness of breath and consumption.

SMALL DAISY
(Bellis minor)

It is good against fevers, inflammations of the liver, inflamed eyes, scrofula and coagulated blood.

DAFFODIL
(Narcissus)

There are several sorts of Daffodils, but they all have the same virtues and therefore I shall not treat particularly of them. They grow in gardens and sometimes wild, flowering generally in April.

They have a hot and dry nature. The roots, pounded with honey are good against burns, bruised sinews, dislocations and old aches. They take away freckles and heal abscesses and sores, and they draw out thorns and splinters. A decoction of the roots is a great emetic.

DANDELION
(Dens leonis)

It grows in fields and meadows, flowering for most of the summer.

It has a cold and dry nature. It strengthens the stomach, causes good digestion, reduces inflammation of the liver and cleanses the kidneys and bladder.

DARNEL GRASS or
Ray-Grass
(Loliumrubrum)

It grows along the borders of fields, and ripens in July and August.

A decoction of it in red wine stops dysentery, diarrhoea and disorders of menstruation and urination.

Daffodil

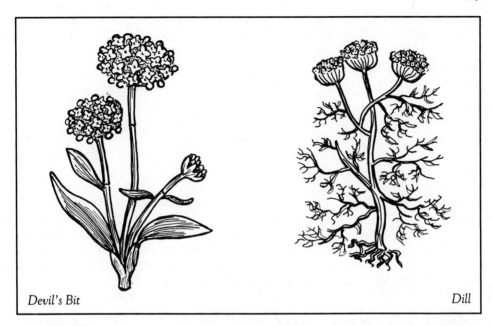

Devil's Bit

Dill

DEVIL'S BIT
(Morsus diaboli, Succisa)

It grows in meadows and pastures, flowering towards the latter part of the summer.

It is an antidote to poison, prevents fevers, dissolves the congealed blood which causes bruises and removes obstructions of the liver and spleen. Added to a bath it is good for old aches, strains and rheumatic pains.

DILL
(Anethum)

It very much resembles common Fennel, but it seldom grows so tall or so much branched. It grows in gardens, flowering in July, and the whole plant has a strong scent.

Dill has a hot dry nature. It expels flatulence, alleviates colic pains, stops vomiting and diarrhoea, and provokes urine. Applied externally it eases pain and heals tumours. The seed is a remedy to stop hiccups and also vomiting.

DITTANDER or
Pepperwort
(Lepidium, Piperitis)

It has long broad serrated leaves; the stalks and branches are round and smooth, growing about 2 ft high, and the flowers are small, white and four-leaved. It grows in moist places and near rivers, flowering in June and July. It has a hot sharp taste.

The roots and leaves are very good against sciatica when mixed with goose-grease and applied to the afflicted area.

When chewed, they draw catarrh from the glands of the throat in great amounts.

SHARP POINTED DOCK
(Lapathum acutum vel oxylapathum)

The broad, round leaved Wild Dock is much of the same nature with this, so I shall not particularly mention it.

The roots, made into an ointment with tobacco, are a great cure for itchy and scabby skin. A decoction of the roots in ale or whey, taken internally is an excellent cure for any scurvy. The seed is useful for stopping all haemorrhages and flows of blood.

GREAT WATER DOCK
(Lapathum maximum aquaticum)

The root is thick and large; the leaves are 2 ft long and not more than four fingers broad, the stalks are large and thick, about 4 or 5 ft high, while the flowers are yellow and set in thick whorls about the branches. It grows in large ponds.

It is very good against scurvy, ulcers of the mouth and gums and against all kinds of flow.

GARDEN DOCK or Patience or Monk's Rhubarb
(Lapathum hortense folio oblongo, Lapathum sativum, Patientia)

This dock frequently grows 5 or 6 ft high; it has long pointed leaves, the stalk is red and the flowers are staminous.

It is somewhat purgative, it removes obstructions and is beneficial for the liver and spleen.

DODDER
(Cuscuta major)

This is a strange herb, having neither leaves nor root. It consists of a number of slender red filaments twisting itself around neighbouring plants and sucking nourishment from them. It bears several white flowers.

It has a hot dry nature, but this nature is changed by the influences of surrounding herbs. It removes obstructions of the liver, spleen and bladder, and is good against old fevers, jaundice and itchy conditions. The Dodder that grows upon Flax, if taken internally, helps to break up gravel.

DODDER OF THYME or Small Dodder
(Epythymum sive cuscuta minor)

This grows on thyme (as opposed to the Greater Dodder which grows on nettles, flax and so on).

It is beneficial for conditions of the spleen and for all eruptions of the skin. It is also a good cleanser.

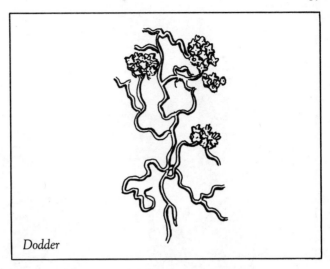

Dodder

DOGBERRY-TREE
or Gatter-Tree
(Cornus)

It is a low tree, or rather a shrub which, being dry, is difficult to cut. It is knotty with many joints, and full of pith. The flowers are white, growing in tufts, being succeeded by small round berries, which are black when ripe. It grows in hedges and under woods, and flowers in April and May while the berries ripen in September.

The fruit is cold, dry and astringent. It is therefore good against diarrhoea and dysentery. It also strengthens weak stomachs. The leaves and tender buds will heal fresh wounds and stop bleeding.

DOG BRYAR or
Wild Rose or Hip-Tree
(Rosa canina,
Cynosbatos)

It grows in hedges, flowering in June, while the fruit ripens in October.

The flowers are very good against excessive menstrual flow. The pulp of the hip or berries is good for disorders of the mouth and chest. The hips, when reduced to powder, and a spoonful taken in white wine, twice or three times, seldom fails to cure the gravel.

GREAT DRAGONS
(Dracontium majus)

The roots are good for asthma, bad coughs, catarrhs, malignant ulcers and eruptions of the skin. The juice of the root, dropped into the eyes, removes cataracts and morbid growths. The green leaves are beneficial when applied to wounds and are excellent to drive away any malignity from the heart.

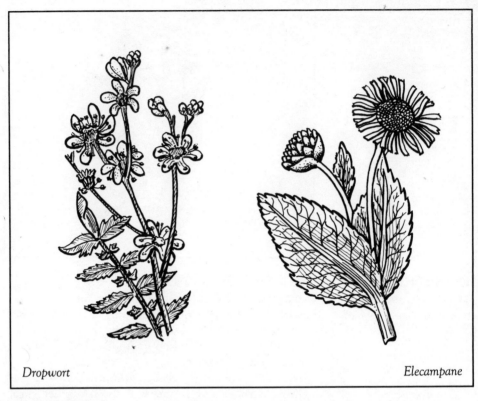

Dropwort   *Elecampane*

DROPWORT
(Filipendula)

It grows in gardens and also wild, flowering in June and July. It provokes urine, cleanses the kidneys and bladder, and expels flatulence. It is also benficial for diseases of the mouth.

WATER DROPWORT or
Hemloc-Dropwort
(Denanthe cicutae facie)

It grows in and near running water. It has a poisonous nature and anyone who eats any quantity of it dies within twenty-four hours.

The roots crushed and mixed with honey is effective in curing abscesses. The leaves pounded and mixed with bay salt cure several disorders of the skin.

DUCKS MEAT
(Lens palustris)

It frequently grows in ponds and ditches, covering the whole surface of the water.

It has a cooling nature and, mixed with wheaten flour or barley meal, it is very beneficial for boils, acute fever, gout and all kinds of inflammations.

DYERS WEED or
Common Would
(*Luteola falicis folio*)

The leaves are long and narrow; the stalks are hollow and channelled, and are about 3 or 4 ft high. The flowers grow in long spokes of a pale yellow colour. It grows commonly in uncultivated places, and flowers in May.

It has a hot dry nature, and is useful for curing both wounds and scrofula.

E

ELDER TREE
(*Sambucus*)

This flowers in May with the berries ripening in September.

The leaves, tender tops and the inner green bark, purge bilious conditions. A small amount of the seed pounded and taken in wine will disperse an accumulation of watery fluid. The green leaves are good against all sorts of inflammations, while the flowers expel wind from the stomach. The berries can be used in gargles for sore mouths and throats.

DWARF-ELDER or
Dane Wort or Wall
Wort
(*Ebulus chamaeacte vel
sambucus humilis*)

It grows in moist fruitful ground, and has much the same nature and virtues of Common Elder.

It is very beneficial in causing watery evacuations.

ELECAMPANE
(*Enula campana,
Helenium*)

It grows wild in meadows, and is often planted in gardens.

It is the root that is principally used in medicine. It provokes urination and menstruation, cures consumptive coughs and shortness of breath, aids digestion, alleviates pains in the joints and, made into an ointment, cures itchy conditions. It is also very good for chest problems, and cures most disorders of the lungs, such as difficulty or suspension of breathing, hurried respiration and asthma.

ELM TREE
(*Ulmus*)

The leaves and inner bard heal and consolidate wounds, bruises and fractured bones. The liquid that is found in the leaves removes freckles, pimples and spreading eruptions. The bark is abstersive and is frequently used in gargles for sore mouths and throats. The inner bark being scraped off and steeped in water for twenty-four hours, is exceeding good to be applied to burns and scalds.

Elm *Eyebright*

GARDEN ENDIVE
(Endiva,
Scariola)

It grows wild and flowers in June.

It comforts and refreshes a weak stomach, stops burning diarrhoea, opens obstructions of the liver and is good against jaundice and burning fevers. Externally applied, it is beneficial for inflammations and abscesses. The leaves mixed with oil of roses and laid against the forehead, alleviates a headache.

ERINGO or Sea Holly
(Eryngium marinum)

It grows near the seaside, flowering in June and July.

It is the root that is principally used. It provokes urination and menstruation, it encourages flatulance and removes obstructions of the liver, kidneys and bladder. It cures colic pains, is good for the nerves and as a general restorative. It is also useful against cramps and convulsions, and is good for consumptive persons.

EYEBRIGHT
(Euphrasia)

It grows in meadows and common pastures, flowering in July and August.

It is good for all disorders of the eyes, the juice being mixed with white wine or distilled water and dropped into the eyes.

Eringo

In short, it generally strengthens the eyesight wonderfully. The pounded leaves cure whitlows.

EYEWORT

It is a small plant like the violet. It bears blue flowers and grows on rocks.

A decoction taken internally is very good against epilepsy, lethargy, apoplexy and migraine, and is also a powerful cleanser.

F

FEATHERFEW
(*Matricaria, Parthenium*)

It is commonly planted in gardens, and it flowers in June and July.

Externally applied, it is good for burning fevers, boils and all bilious inflammations. It is useful for most disorders of the womb. It is also good for all kinds of acute fever, especially if 2 oz of the juice is taken before a fit begins. A poultice made with featherfew, rue, oaten meal, milk and rusty bacon is extraordinarily good against boils, in order to break them and afterwards to draw the corruption out of them.

FELIX WEED
(Sophia chirurgorum)

The leaves are jagged and somewhat hairy; the stalks are round and hard, about 2 ft high, while the flowers are small, yellow and four-leaved. It grows frequently in sandy ground and among rubbish, and it flowers in June.

The seed is excellent for dysentery and for all flows of blood; if pounded and applied to all ulcers and sores, it heals and conglutinates them. It provokes urine, and is good for stone and gravel.

COMMON FENNEL
(Foeniculum vulgare)

It is planted in gardens and flowers in June.

The leaves and seed increase milk in nurses. The roots are very good against stone and gravel, and the whole plant is extraordinarily good in opening obstructions of the liver, spleen and lungs. The seed strengthens the stomach and prevents vomiting. It is a powerful optic medicine, excellent for most disorders which the eyes are subject to, for it strengthens them wonderfully. The leaves pounded with vinegar are very good against boils and other inflammations.

HOGS FENNEL or Sulpher Wort or Sow Fennel
(Peucedanum)

This herb has a weak slender stalk, the leaves being larger than the leaves of Fennel. It bears five-leaved small yellow flowers. It grows near the seashore, and flowers in July.

It provokes urination and menstruation and removes obstructions of the kidneys. It is good against coughs and shortness of breath. The juice, mixed with egg and applied with oil of roses and vinegar, eases an old headache and is useful in an epilepsy. Taken into the nostrils, it is good against apoplexy and lethargy.

FENUGREEK
(Foenum graecum)

It is planted in gardens, and the seed is the only part used.

It causes abscesses, boils and all kinds of tumours to suppurate. It is also excellent in curing bites of mad dogs.

FEMALE FERN or Brake
(Filix foemina vulgaris)

It grows in woods and mountains.

The root crushed and $\frac{1}{2}$ oz of it taken with honey and water will expel worms from the body. A decoction of it in wine will remove obstructions of the liver and spleen. Around midsummer, the country people burn the stalks and leaves of it in order to make ashes with which to whiten their linen clothes.

MALE FERN
(Filix mas vulgaris)

It grows in hedges and shady places, and has the same virtues which are attributed to the female fern.

The roots, which are the only part used in medicine, open obstructions of the liver and spleen and are good in curing rickets in children.

WATER FERN or Flowering Fern or Osmund Royal
(Fixil florida, Osmunda regalis)

This is the largest of all the ferns, and not much unlike the female Fern, only that is is not indented about the edges. It grows in woods and marshy boggy places.

The roots are good against bruises, dislocations and ruptures. It is sometimes put into healing plasters.

FIG TREE
(Ficus)

The milk or juice of the figs is good against freckles, spreading sores, itching and roughness of the skin. A little cotton dipped in it will ease a toothache. The ashes of the tree mixed with oil of roses will cure burns. Figs break up stone of the bladder and help heal lung disorders. When applied as a poultice, they cure scrofulous tumours and extract broken bones.

FIGWORT
(Scrophularia major)

It has square brown fistulous stalks, about 3 ft high; the leaves are indented at the edges like Nettle leaves and they smell like elder. The dark purplish flowers grow in small clusters. It grows along the borders of fields, under hedges and about lakes and ditches, and flowers in June.

It dissolves all kinds of hard swellings and heals ulcers and cankers if it is pounded and applied with salt. Washing with its juice takes away redness of face and it is good against scrofulous tumours and boils.

WATER FIGWORT
(Scrophularia aquatica)

It is larger and taller than figwort, and the root has no knots or tubercles. It grows in moist places and by ditch sides, flowering in June.

The virtues of figwort may be attributed to this. It is useful in healing wounds and itchy conditions.

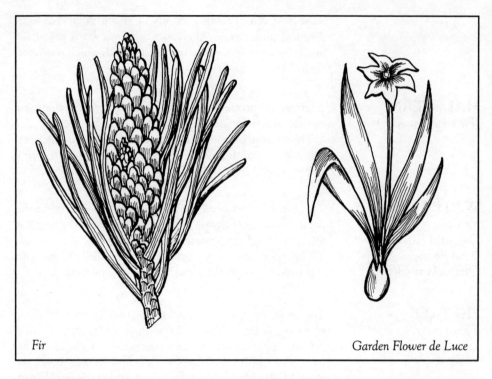

Fir   *Garden Flower de Luce*

FIR TREE or Male Fir Tree
(Abies mas)

The leaves and top of this tree can be used against scurvy. The turpentine or liquid resin that can be extracted from the tree is somewhat purgative, and also provokes urine and is beneficial to the bladder, kidneys and arthritic conditions. It is good for wounds, being healing and cleansing.

YELLOW WATER FLAG or Bastard Acorus or Water Flower de Luce
(Acorus adulterinus, Pseudoacorus, Gladeolus luteus, Iris palustris lutea)

The roots have a cold dry nature, and have great binding qualities. A decoction made from them stops dystentery, diarrhoea, haemorrhages and excessive menstrual flow.

MANURED FLAX
(Linum sativum)

Linseed softens and breaks abscesses and tumours. It takes away spots and freckles from the face. A decoction strengthens and clears the eyes. Taken in suppositories it eases colic pains, and taken as a paste, it is good for all disorders of the lungs. Linseed is very good when applied to burns or scalds.

PURGING FLAX or Dwarf Wild Flax or Millmountain
(Linum cartharticum herba minuta)

This is but a small plant, seldom growing above a span high. It has round slender stalks, having two oblong small leaves at each joint; it has many branches, bearing several five-leaved white flowers, which are succeeded by seed vessels, which are in the shape of the Common Flax, but much smaller. It grows on dry ditches and hilly places, flowering in June and July.

When boiled in ale and drunk, it is a powerful purgative and the common people frequently make use of it.

WILD NARROW LEAVED FLAX
(Linum sylvestre)

The leaves are thinly set and the flowers are blue.

It provokes urine, breaks up stone and opens obstructions of the liver and spleen. It is particularly good for curing jaundice.

FLEA BANE
(Conyza vel pulicaria minor flore globoso)

The stalks are hard and of a reddish-brown colour. The leaves are small and woolly, and about 1 in long. The flowers are yellow like the flowers of Tansy. It grows in moist places, flowering in August and September.

It provokes urination and menstruation, and expels a still born child. It is good against jaundice, colic pains, epilepsy, wounds and swellings. It is known as Pulicaria because it destroys fleas.

GARDEN FLOWER DE LUCE
(Iris nostras hortensis)

This flowers in May and June.

The roots encourage sneezing and vomiting; they provoke menstruation and they eliminate bile and cleanse the lungs, clearing them of all obstructions.

FLOWER GENTLE
(Amaranthus ceu flos amoris)

The stalks are large and channelled, about 3 or 4 ft high, the leaves are long and broad, and of a light green colour, and on the tops of the stalks grow long spikes or deep red staminous flowers. It is planted in gardens, flowering in July and August.

The flowers stop haemorrhages and flows of all kinds.

FLUELLIN or Male Speedwell
(Veronicamas supina vulgatissima)

The stalks generally lie on the ground, shooting out fibres at the lower joints. The leaves are oval, hairy, notched about the edges and of a pale green colour, while the flowers grow in short spikes, each with one small bluish purple leaf, cut into four parts. It grows in woods and shady places, flowering in June.

It is good against all obstructions of the kidneys; it heals wounds, cleanses the blood, and is good against all sorts of flows and haemorrhages.

FLUELLIN or Female Speedwell
(Elatine, Veronica faemina)

It is a long low plant creeping on the ground. It has slender hairy stems, bearing large, soft and hairy leaves. The flowers have the same shape as the flowers of Toad-Flax or Larkspur. It grows in cornfields and moist places.

It has the same virtues of the former, only this herb is weaker in operation.

FOOL STONES or Dog Stones or Male Satyrion or Standergrass
(Satyrium mas, Orchis)

It bears five or six smooth long shiny leaves which are almost like lily leaves, only that these are black spotted. The flowers grow in spikes and have a purple colour. It has two oval roots, which are nearly as big as small olives, and which are of a whitish colour. It grows in moist meadows, and flowers in May.

Fluellin

Fool Stones

A decoction of the roots drunk in goat's milk excites sexual desire, aids conception and strengthens the genitals. The roots restore the strength of someone suffering from consumptive fever. They also stop diarrhoea, and if externally applied, they purify rotten ulcers. It is said that if men eat the largest roots, they will beget sons, and if women eat the smallest roots they shall bring forth daughters.

FEMALE FOOL STONES or Female Satyrion
(Satyrium faemina, Orchis faemina, Morio faemina)

This is smaller than the former herb, having no spots on the leaves. The flowers are also smaller and some of these are deep violet, some white and some of a carnation colour. It grows in the same places with the former, but flowering somewhat later.

It has the same properties as the former.

Foxglove

FOXGLOVE
(Digitalis)

It grows in mountains, ditches and highways, flowering in June and July.

A decoction drunk dissolves viscous, clammy and slimy phlegm. It opens obstructions of the liver and spleen. It is a great purgative and it encourages vomiting. It is good for any obstructions of the lungs and also for epilepsy. Made into a poultice with hogs lard, it cures scrofula.

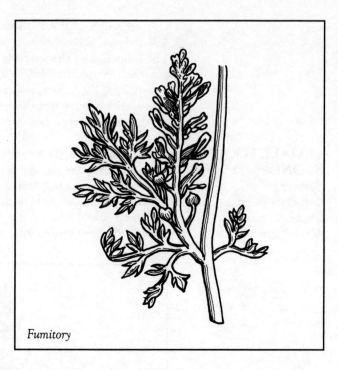

Fumitory

FUMITORY
(Fumaria)

It has square stalks with many branches; the leaves are weak, tender and finely divided. The flowers grow in spikes and are purple above and white underneath. The whole plant has a bitter taste. It grows along the borders of fields and tilled grounds, among wheat and barley, flowering in May.

The juice dropped into the eyes sharpens the sight. Taken internally, it cleanses the blood; externally applied, it is good against all eruptions of the skin.

FURZ
(Genista spinosa)

The leaves have a hot dry nature, and are therefore astringent. A decoction prevents excessive menstrual flow and also diarrhoea. The seed is good for jaundice and for killing worms.

G

GARLICK, manured
(Allium sativum)

It is a great antidote to poison, and good for diseases of the mouth and lungs. It has a hot, dry nature and Galen, the Prince of Physicians, called it the Poor Man's Treacle. It is good for coughs, asthma and bowel conditions and for expelling flatulence. Pounded with vinegar, it eases toothache, headache and dissolves hard swellings.

To cure any cough, asthma or shortness of breath, boil a handful in 2 quarts of spring water. Allow 1 quart to boil away and strain it, then add 1 quart of honey to make a syrup, then take a spoonful of this at a time.

CROW-GARLICK
or Wild Garlick
*(Allium sylvestre
tenuifolium)*

Instead of leaves it has long round small and hollow blades, from which grows up a round hard stem, about 2 or 3 ft long; upon this grows the flower and seed. It grows in pasture fields, hedges and meadows, flowering in June.

The roots are good against scalds, itches, leprosy and swellings. They provoke menstruation and eliminate the afterbirth. If they are infused in white wine overnight and drunk in the morning before breaking the fast, they give relief from the pain of stone and gravel.

Take the roots and blades of this garlic, let them be well pounded, then strain 1 quart of juice out of them, then add a $\frac{1}{4}$ oz of pounded aromatic cloves. This will infallibly cure rheumatism or rheumatic pains if the patient is well rubbed with it twice daily for three days, in the morning before he arises and at night when he retires. However, he must ensure that he keeps himself warm for an hour after the operation is performed, likewise, he must not stir out of his chamber during the said three days lest (his pores being open), he should take a cold.

BOG GAUL or Sweet
Willow or Dutch
Myrtle or Wild Sumac
*(Myrtus brabantica,
Eleagnus cordi,
Rhus sylvestris)*

It has a strong smell and a bitter taste. It grows in bogs, low land, meadows and near rivers, and flowers in May and June.

It has a hot dry nature. It is useful in destroying vermin and the common people sometimes put it on their drink instead of hops.

BASTARD GENTIAN or Dwarf Felwort
(Gentianella)

A decoction of it drank encourages vomiting and clears the lungs. It is generally good against infection and, when applied externally, it heals wounds and ulcers.

GERMANDER
(Chamaedrys, Trissago)

This grows in gardens, flowering in June and July.

It opens obstructions of any part of the body, especially the liver and spleen. It is good for coughs and the loosening of phlegm. It provokes urination and menstruation.

WATER GERMANDER
(Scordium)

It has square downy stalks about 1 ft high; the leaves are round, wrinkled, soft and whitish, and serrated about the edges. The flowers grow in thin whorls, and are of a reddish colour. It has a strong aromatic smell. It grows in moist meadows and fens, and flowers in June.

It is very good in provoking urine. It removes obstructions of the liver, spleen, kidneys, bladder and womb.

WILD GERMANDER
(Chamaedrys sylvestris)

It usually grows in rich soil, and is very like Garden Germander both in shape and qualities, but is more effective.

It is good against arthritis, rheumatism and gout. It

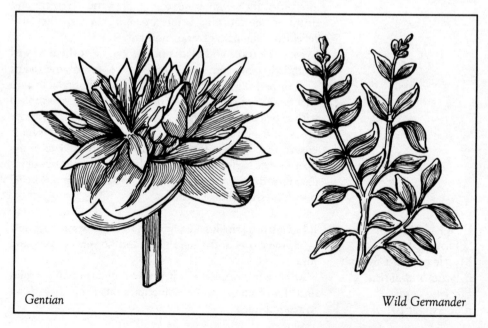

Gentian

Wild Germander

Gladwyn

removes pains in the joints, and opens obstructions in any part of the body. It is very good for convulsions in children, either taken internally or worn around the wrist or neck.

SEA-GIRDLE
(Fucus phasganoides)

It has long narrow leaves; the root is rough and thick and full of fibres, and it grows on rocks in the sea.

It is good against boils, inflammatory tumours and gout.

STINKING GLADWYN
(Spatula foetida, Xyris)

It bears long narrow sharp-pointed leaves of a dark green colour. The stalks are smooth and round, on which grow flowers, which are like the flower De Luce, but which are smaller and of an ash colour. It grows in stony places, hedges and borders of woods, flowering in June.

The seed, taken internally, provokes urine and cures obstructions and hardness of the spleen. It is a remedy for scrofula and scrofulous swellings. It also kills the stinking insects, called bugs, and which were of late years brought to this Kingdom, if the place where they frequent be rubbed with the juice.

GLASSWORT or
Saltwort
(*Kali spinosum*)

It has thick brittle stalks about 1 ft high; among the leaves, which are long, sharp-pointed with prickly tops, grow small yellow flowers. It grows near the sea shore and has a dry nature.

From it is made Sal alkali, which is an ingredient of the finest glass. A kind of fine salt can be extracted from its ashes. It is purgative and removes all kinds of obstructions.

YELLOW GOATS BEARD or Go to Bed at Noon
(*Tragopon luteum*)

This plant grows near running water and in pasture ground.

The roots are restorative for consumption. They break up stone and are good for asthma and painful urination.

GOATS-RUE
(*Gaelega*,
Ruta capraria)

It grows to a height of 3 ft, and has blue flowers growing in long spikes which hang downwards. It is planted in gardens and grows wild in rocky places, and flowers in June and July.

It is very good against poisons and malignities, for it expels the poison through the skin by inducing perspiration. It will kill and eliminate worms from children, either by drinking the juice or if the juice is boiled in linseed oil and applied to the child's navel. A spoonful of the juice of this herb is good

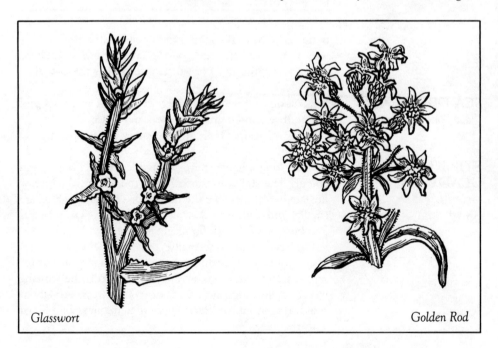

Glasswort   Golden Rod

to be given to children every morning to drink, against convulsions.

GOLDEN ROD
(Virga aurea)

It has long broad leaves spreading on the ground near the root and indented about the edges. The stalks are round, hairy and full of fungous pith, growing to about 3 ft high, with the flowers growing in small yellow spikes. It is usually planted in gardens, but it is also found growing wild. It flowers in July.

It helps to break up stone in the kidneys and expel it in the urine; it clears obstructions of the liver and spleen. It heals all curable wounds, and is used in gargles for ulcers of the mouth and throat.

GOOSEBERRY BUSH
(Grossularia)

The unripe berries stop diarrhoea and flows of blood, and are very good in applications to boils and inflammations. Eaten with meat, they provoke the appetite, and cool the heat of the stomach and liver. The leaves provoke urine and break up stone in the kidneys.

GOOSE TONGUE
(Lingua anserina)

It grows generally in gardens and is good against fainting.

WILD GOOSE TONGUE
(Lingua anserina sylvestris)

The stalks are small, about 1 ft high; the leaves are small, rough and somewhat indented and the flowers are yellow.

It is a remedy for epilepsy and hysterical fits and is useful for all kinds of convulsions.

GOURD
(Concurbita major sessilis)

This plant is grown yearly from seed in gardens, flowering in July. The fruit, which ripens in September, is very large and in the shape of a bottle which will hold several quarts. The leaves are large, rough and woolly, and the flowers are white.

The juice is good against boils, inflammations and burning fever.

SEA GRAPE or Jointed Glass-Wort
(Salicornia, Kali genticulatum)

This plant has stalks without leaves, divided into several branches with knots, each as big as grains of wheat and easily pulled off. It has a salt taste and is full of juice. It grows near the sea.

It has a very dry nature, and is pickled like Samphire to create an appetite.

Gromwell

GRASS OF PARNASSUS
(Gramen parnassi vel flos hepaticus)

It has round leaves like the leaves of ivy, only smaller, among which spring up two or three small stalks about 1 ft high, bearing fair white flowers. It grows in moist places, flowering in July.

It strengthens a weak stomach, stops diarrhoea, provokes urine and expels gravel. It also stops the bleeding of wounds and heals them.

GROMWELL or Gromill
(Lithospermum seu milium solis)

It has long slender hairy stalks; the leaves are brown, rough, oblong and sharp-pointed, and the flowers, which are cut into five segments, are succeeded by white, shining hard seeds, like pearls. It grows in fields, hedges and rough stony places, and flowers in May.

Gromwell has a hot dry nature. The seed, if pounded and drunk in white wine, breaks up the stone in the bladder, expels gravel from the kidneys and provokes urine. Two drams of it taken internally, is an excellent remedy to facilitate labour. It is used in baths for strains, sciatica, and rheumatic pains.

GROUND PINE
(Chamaepitys, Ivy arthritica)

This is a small tender plant creeping on the ground. It has many crooked stalks, the leaves are small, narrow and hairy and the yellow flowers are labiated. It grows in fallow fields and stony ground, flowering in June and July, and has a strong scent.

A decoction in white wine, drunk for seven mornings, cures jaundice, and every day for a month, cures sciatica. It clears obstructions of the liver and spleen, provokes urine and cures wounds, ulcers and lumps of the breast. Made into a poultice with white bread, sweet milk, nerve oil and saffron, it alleviates rheumatic pains. It opens obstructions of the womb, and promotes menstrual flow.

GROUNDSEL
(Erigeron vel senicio)

It grows on banks, walls and among rubbish, flowering in June and July.

It has a cooling digestive nature. The leaves and flowers crushed with wine are beneficial in applications for

Ground Pine

inflammations. The down of the flowers crushed with a little salt, heals scrofula, while the juice taken in ale encourages vomiting. It helps cure jaundice and kills worms.

H

HARE-BELLS or English Hyacinth
(Hyacinthus anglicus)

If the roots are boiled in wine and taken internally, they stop diarrhoea and dysentery and provoke urine. These disorders are more effectually cured by the seed. The roots drank in wine are very good against epilepsy.

HARES FOOT
Trefoil
(Lagopus ceu pes leporinus)

It has small narrow hairy leaves like clover, and rough round stalks, on the tops of which grow small purple flowers. It is found among corn and in fallow fields, flowering in June and July.

It has a dry nature. A decoction of it drunk in wine stops diarrhoea, dysentery and excessive menstrual flow.

HART'S TONGUE
(Lingua cervina vel phyllitis)

It bears long narrow leaves about a span or two long and 2 in wide and their upper parts are smooth and plain. The seeds grows on the underside of the leaves. It grows in shady places, around fountains and stony moist places.

It is dry and astringent. If a decoction is drunk, it will stop diarrhoea and dysentery. It is also good for coughs, consumption and in opening obstructions. It is very good for hysteric and convulsive fits.

GREAT HAWK-WEED
(Hieracium magnum)

The stalks are reddish and rough, the leaves are long and jagged having sharp prickles almost like the leaves of milk thistle. It bears double yellow flowers like those of dandelion, except smaller. It grows in untilled places, in ditches and along the borders of corn fields, and spends most of the summer in flower.

This herb is, in virtue and operation, much like Sowthistle, and is used in the same manner. It is cold and dry and the juice is good for strengthening the eyes.

HAW-THORN or
White-Thorn
*(Spina alba,
Oxyacanthus)*

This tree grows almost everywhere in hedges, flowering in May, with the haws ripening in September. The fruit is dry and astringent.

It stops flows and excessive menstruation. The flowers are very good for breaking up stone in the kidneys and bladder.

HAZEL-NUT Tree
*(Avellana vel corylus
sylvestris)*

Hazelnuts provide little nourishment, are hard to digest, cause shortness of breath, and if a lot are eaten, encourage flatulence. A decoction in mead cures an old cough. If roasted and taken with a little pepper, they are beneficial for catarrh and runny nose or eyes. If burnt and applied with hogs lard, they cure a scald.

HEARTS-EASE or
Three Faces under a
Hood
*(Pansies,
Viola tricolor)*

This grows in gardens, flowering for most of the summer.

The flowers cure convulsions in children, cleanse the lungs and breast and are very good for fevers, internal inflammations and wounds.

COMMON HEATH
or Ling
(Erica vulgaris)

It grows on barren mountains and flowers about midsummer.

It has a dry nature. The juice of the leaves dropped into the eyes strengthens and cures them if inflamed. About a half pint of a decoction in spring water drunk warm, is very good against stone in the bladder if taken for thirty days. After which, the patient must take a bath made of a decoction of it, and this process must be often repeated. The bark is beneficial for diseases of the spleen.

**HEDGE-BERRY
TREE** or the Wild
Cluster Cherry or
Bird's Cherry
*(Cerasus avium nigra,
Racemosa)*

It has a binding quality, it strengthens the stomach, and stops diarrhoea.

HEDGE-MUSTARD
*(Erysimum vulgare sive
irio)*

It grows everywhere along waysides, and in untilled stony places, flowering in June and July.

A syrup made from the seed and some honey loosens viscous and clammy phlegm. It is also good against an old cough, shortness of breath, sciatica, jaundice and colic pains. A decoction in wine is very good for cholic.

HELLEBORE or Great Bastard Black Hellebore or Bears-Foot or Sellerwort
(Helleboraster maximus, Contiligo, Helleborus nigra foetidus)

It grows on mountains, flowering in early spring.

The roots are cleansing. Thrust into the ears of cattle, it helps them against the diseases of the lungs, for it draws the malignity from them into the ears, which is soon after evacuated. Taken inwardly, it cures jaundice; applied externally, it cures abscesses and sores, leprosy, itches and eruptions of the skin.

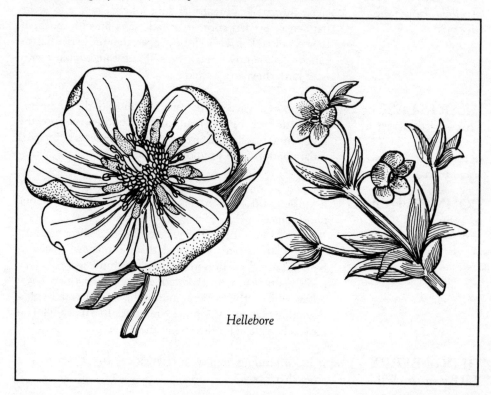

Hellebore

HELM or English Seamatweed or Marram
(Spartum anglicanum ceu gramen sparteum)

The stalks are woody, on which grow few sharp-pointed leaves, and the flowers, which appear yellow and not much unlike the flowers of common Broom, appear in June.

It has a hot dry nature. The flowers and seed provoke vomiting, but not to a dangerous extent. The juice, taken while fasting is beneficial for sciatica and inflammation of the throat.

HELMET FLOWER
(Napellus)

It is a small tender plant, having spotted round leaves. The stalk rises about 1 ft high, on which grows a double flower, divided into two parts, one yellow and the other blue.

It helps children with rickets, and is also good for a prolapsed womb.

HEMLOCK or the Ordinary Great Hemlock (*Cicuta major*)

It has a very cold nature, and is therefore good against boils, inflammations, burning fever, the hardness of the liver and the spleen. It also has a poisonous nature – the best cure, if eaten accidentally, is to drink plenty of good old wine.

LESSER HEMLOCK or Fools Parsley (*Cicutaria tenuifolia*)

It has spotted stalks and grows in moist rich soil and in kitchen gardens.

The root has a hot and dry nature. It provokes menstruation and assists births and expels the afterbirth. It is also beneficial for chest problems and is generally strenthening.

WATER HEMLOCK (*Cicutaria palustris, Phellandryum*)

It has a thick jointed stalk, shorter than that of ordinary hemlock. It has large-winged leaves, finer and more tender than common hemlock, and the flowers are white, but with a red cast. It grows in running waters, ditches and ponds, flowering in June.

It is thought to be similar in virtues to Common Hemlock, but more poisonous.

HEMP or Manured Male Hemp (*Cannabis mas sativa*)

It has a hot dry nature. The seed encourages flatulence, and, pounded and taken in white wine, cures jaundice. It is good for removing obstructions of the liver and easing old coughs. The juice of the green leaves, dropped into the ears, alleviates their pain and expels all kinds of vermin out of them. A decoction of the roots helps to cure contractions of the sinews, and eases gout.

HENBANE or Common Black Henbane (*Hyoseyamus nigra*)

The stalks are large, round and soft, growing 2 or 3 ft high. The leaves are broad, downy and greyish, and the flowers are bell-shaped, have but a single petal, and are of a pale yellow colour, full of purple veins. It grows near highways and paths, and in sandy ground.

It is good against all kinds of inflammations, if the leaves or juice are applied. The seed is beneficial for coughs, catarrh, spitting of blood and excessive flows of blood. It alleviates the pain of gout and the swelling of women's breasts after delivery of a child.

HENBIT or Great Henbit
(Altine haederula major)

It has several upright round knotty stalks; the leaves are not unlike Pellitory of the Wall, but are longer and less hairy, and it bears small white flowers, deeply cut. It grows in moist shady places, under hedges and bushes. It flowers about midsummer.

It is good for inflammations and ulcerations of the eyes. If the juice is dropped into them, it is beneficial for pains in the ears.

SMALL HENBIT or Ivy Chick-Weed
(Alsine haederacea minor)

It is similar to the Greater Henbit, except that it is smaller and spreads on the ground rather than growing upwards.

A decoction is a powerful remedy against the heat of scurvy, and the itch of the hands, if they are bathed in it.

HERB ROBERT
(Geranium robertianum vel gratia dei)

It has round, tender, hairy reddish stalks, full of joints. The leaves resemble chervil leaves, and the flowers, which grow at the joints, are of a clear purple colour, inclining to red and are succeeded by heads which resemble the head and beak of a crane; it is a kind of Cranes-Bill. It grows in stony places, old walls and under hedges.

It stops the bleeding of wounds if crushed into an application. A decoction is good against rotten ulcers and sore mouths, and it is particularly good for scrofula.

HOLLY-TREE
(Agrifolium)

The berries have a hot nature, and it is believed that five of them when eaten relieve cholic and act as a purgative. The bark is sometimes made into birdlime.

HOLLYHOCKS
(Malva arborea seu hortensis)

It grows in gardens, flowering in July and August.

The leaves eaten with a little salt, help to ease the pains and ulcerations of the kidneys. It breaks up the stone of the bladder and kidneys. It is also used in gargles for swelling of the tonsils and to relax the uvula.

HONEY-TREE
(Melianthus)

It can be grown in gardens with very little cultivation and industry. It bears a yellow flower, which is succeeded by a lump of congealed honey.

HONEY-SUCKLE or Woodbind
(Caprifolium, Matrisylve, Percilymenum)

It grows in hedges, flowering in June and July.

This has a hot dry nature, and the fruit cures obstructions and hardness of the spleen, if it is drunk in wine for forty days. It is also beneficial for coughs, shortness of breath, and assists women in hard labour. The leaves are used in gargles for sore throats and the flowers for cramps and convulsions. The leaves and fruit however, are dangerous for pregnant women. A powder made from the leaves is exceeding good for the ague, taking as much of it as can be put on a shilling in a glass of white wine, before the fit comes on.

HOPS Manured
(Lupulus sativus)

The flowers are of a hot, dry nature. They open obstructions of the liver, spleen and kidneys, and purge the blood through the urine. They are also good for those who are troubled with the itch, or such like infirmities. The juice dropped into the ears, cleanses them.

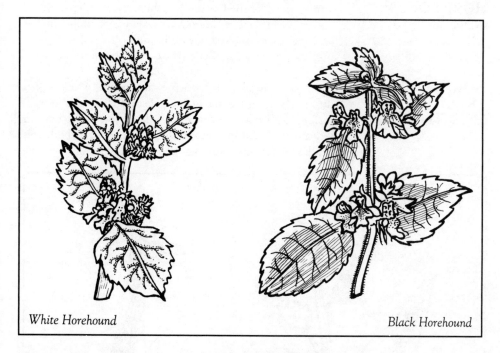

White Horehound

Black Horehound

HOREHOUND or
White Horehound
(Marubium album ceu prassium)

It grows in gardens, flowering in June.

It removes all obstructions of the liver and spleen. A syrup made from it is very effective in curing old coughs, tuberculosis, and ulcerations of the lungs. A decoction in wine opens obstructions of the womb, encourages menstruation, expels the afterbirth and stillborn child and assists women in hard labour. Pain of the ears is relieved by the juice being dropped into them. A decoction of the leaves cure jaundice.

HOREHOUND or Black or Stinking Horehound
(Marubium nigrum vel ballote)

It is taller and more branched than the White Horehound. The stalks are square, and hairy but of a blackish colour, the leaves are larger, longer and darker, resembling dead Nettle leaves. The flowers grow whorl-fashion round about the stalks, and it grows in untilled places, near highways, in hedges and borders of fields, flowering in July.

Pounded with salt it can be applied to those bitten by a mad dog, while the leaves, roasted in a cale leaf, repel piles. It can be applied with honey to rotten ulcers. Taken inwardly, it is good against fits, and is useful against poison.

WATER HOREHOUND
(Marubium aquaticum)

The leaves are of a dark green colour, hairy and somewhat wrinkled, while the flowers are white and smaller than the flowers of other Horehounds. It grows near moist ditches and water courses, and flowers in July.

It has properties similar to the other Horehounds.

HORN BEAM TREE
(Betulus sive carpinus)

The roots infused with wine are good, when taken internally against pains of the side. The fruit mixed with grease, heals burns and scalds. A decoction of the bark in vinegar, is useful in preventing toothache, and the leaves can be used against swellings and inflammations.

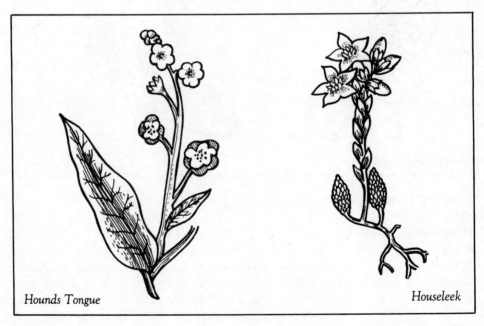

Hounds Tongue

Houseleek

HORSETAIL or the
Great Marsh Horsetail
*(Cauda equina vel
equisetum)*

It grows in ditches and marshy grounds.

A decoction of it taken internally stops excessive menstruation, dysentery and all flows of blood. If it is pounded or crushed and applied to wounds, it stops the bleeding and any further inflammation. A decoction of the whole plant is good against a cough and difficulty of breathing. The juice, if sniffed into the nose, will stop a nosebleed.

**NAKED
HORSETAIL** or
Shavegrass
*(Equisetum folis nudum
non ramosum ceu
juncum)*

It grows in fenny and marshy ground, and its virtues are as the Stinking Water Horsetail.

**STINKING
WATER
HORSETAIL**
(Equisetum foetidum)

This species grows in muddy pools, dirty ditches and drains. Both this kind and the former have much the same qualities as the Great Horsetail.

**HOUNDS
TONGUE**
(Cynoglossum vulgare)

It has a hard rough brown stalk, about 2 or 3 ft high; the leaves are long, much like the leaves of Garden Bugloss, but narrower, smaller and not so rough, being covered with a fine down like velvet. It grows in sandy ground, under hedges and near paths and highways, and it flowers in June and July.

The root is very good for wounds, ulcers, inflammations, discharges of the lungs, haemorrhages and gonorrhoea.

HOUSELEEK or the
Great Houseleek
*(Sedum majus vulgare vel
sempervivum majus)*

It grows on old walls and thatched houses, flowering in July and August.

It has a cold dry nature. A decoction of its juice taken internally is good for dysentery and diarrhoea. Mixed with barley meal and oil of roses, it is effective in easing a headache, while the juice dropped into the eyes is good for inflammations. It is also useful for burns, scalds, inflammations, burning fever and the gout.

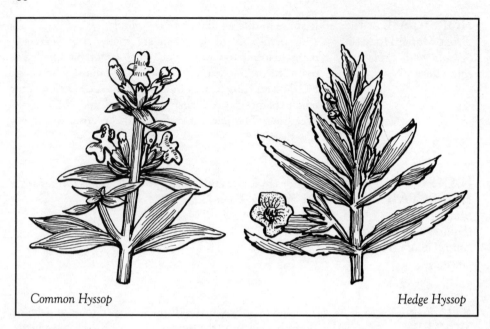

Common Hyssop

Hedge Hyssop

HYSSOP or Common Garden Hyssop *(Hyssopus)*

It is planted in gardens, flowering in June and July.

It is very beneficial for moist asthma, and will open phlegmatic obstructions of the lungs. A decoction in vinegar, if held in the mouth, eases toothache. It dissolves congealed blood, black and blue marks and cures itchiness, scurf and mange, if the skin is washed therein. It is also good for diseases of the head and nerves. It is specifically good for disorders of the lungs.

HEDGE HYSSOP *(Gratiola)*

It has square stalks scarcely 1 ft high, the leaves are narrow and sharp-pointed and the flowers, similar to those of the Foxglove in shape, have a pale yellow colour. It grows in sandy hills, and flowers in July.

It is a powerful purgative and it helps to eliminate clammy phlegm that accumulates in the lungs.

I & J

IRON-WORT or Narrow Leaved Allheal
(Sideritis arvensis rubra)

The stalks are hard and square, the leaves long and the flowers red.

It is very useful in treating all wounds and ulcers. It effectively prevents the spitting and pissing of blood.

IVY-TREE
(Haedera arborea corymbifera)

It flowers in June, with the fruit ripening in winter.

The juice of the leaves cures wounds, ulcers, burns and scalds and, when applied with oil of roses, it cures a headache. Sniffed into the nose it purges and cleanses the head. A decoction of the flowers in wine drunk twice a day, stops dysentery and the gum removes spots and freckles.

JACK BY THE HEDGE or Sauce Alone
(Alliaria)

This herb, when it first springs up has roundish leaves, almost like the leaves of March Violets, but much larger. The stalks are about 2 ft high, on the tops of which grow many white flowers. It grows about the borders of meadows, in moist pastures and in hedges, flowering in May.

It has a hot dry nature. It provokes urine, encourages sweating, is good for the lungs and is an effective antidote to poison. It is also outwardly applied with good success against gangrene.

JASMINE or Jessamy
(Jasminum)

It is planted in gardens, and flowers in June and July.

It has a hot dry nature. It cures red spots and pimples, and dissolves swellings and lumps on the skin. It also clears catarrh, and facilitates the process of birth.

JERUSALEM ARTICHOKE
(Flos solis pyramidalis)

It grows in gardens, and is effective in opening vessels, provoking urine and increasing male fertility. It has much the same properties as potatoes or common artichokes.

JEWS-EAR
*(Fungus sambucinus,
Auricula judae)*

This is a fungus which frequently grows on the trunk of the elder-tree, being wrinkled and turned up like an ear, whitish on the outside and black inside, having several little veins.

Made into a powder, it cures chilblains, dissolves swellings, cures inflammations of the tonsils and throat.

ST JOHNS-WORT
(Hypericum)

It has a brown red stalk about 2 ft high; the leaves are long and narrow, which if held against the light, appear full of small holes (therefore it is known as Perforata) and the flowers are yellow. It grows near highways, along the borders of fields, in hedges and among bushes. It flowers in June and July.

It has a hot dry nature. It provokes urine and breaks up stone in the bladder. It stops diarrhoea and cures fevers. The seed, boiled and drunk for forty days, heals sciatica, and if pounded, it makes an effective application to burns, wounds and ulcers.

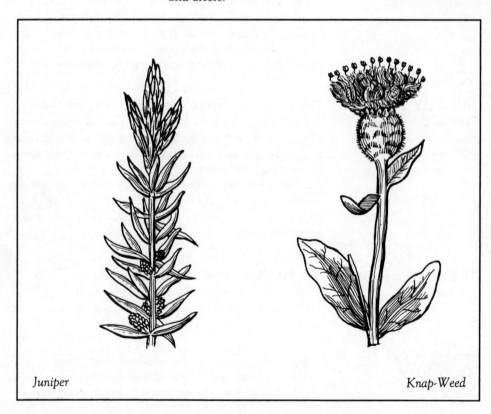

Juniper

Knap-Weed

WHITE STOCK
JULY FLOWER
(Lucoium album)

It grows in gardens, and flowers most of the summer.

It provokes urination and menstruation and expels the afterbirth, if a decoction of the flowers is drunk. If the flowers are made into a plaster with oil and wax, they can be effectively applied to ulcers and inflammations. The juice, dropped into the eyes, strengthens and clears them.

JUNIPER-TREE
(Juniperus vulgaris)

It grows on rocky mountains and in shady places.

The berries provoke urine, and cure old coughs, flatulence and colic pains. The gum of the tree expels worms from the body and stops excessive menstrual flow.

K

KINGS-SPEAR
(Asphodelus verus luteus vel hasta regia)

This herb bears narrow sharp-pointed leaves like a swordblade; the stalks are round, and the flowers grow in yellow spikes. It is planted in gardens but it is sometimes found growing wild near rivers. It flowers in May and June.

The root provokes urination and menstruation. A decoction boiled in the lees of wine and taken internally is effective when applied to sores, ulcers and abscesses. An oil made from it cures burns and chilblains.

KNAP-WEED
(Jacaea nigra vulgaris)

This herb grows to be about 2 ft high; the leaves are long, narrow and bluntly cut about the edges. On the tops of the round stalks grow small round buttons, like those of the blue bottle, from which emerge purple hairy flowers. It grows in meadows and pastures, flowering in June and July.

It has a hot dry nature. It cures abscesses of the mouth and throat if gargled.

KNAWELL or
Parsley Piert or Parsley
Break Stone
(Percepier anglorum vel alchimilla minima montana)

It bears many small narrow hairy leaves, and the flowers are small, growing in clusters at the joints. It grows in sandy places and in fallow fields.

It provokes urine and breaks up kidney stone.

KNOT-GRASS
*(Polygonum mas
centinodia,
Gramen nodosum)*

It has a lot of slender stalks full of joints and creeping on the ground. The leaves are oval and sharp-pointed, and between the leaves and joints grow small staminous flowers, sometimes white, sometimes of a carnation colour. It grows in streets and by waysides, flowering in summer.

It is very good against all haemorrhages, spitting and pissing of blood, diarrhoea, dysentery and excessive menstrual flow. Nosebleeds can be stopped by some juice being placed into the nose. Applied externally the leaves are good for boils and inflammations.

L

LADIES MANTLE
or Lyons-Foot
(Alchimilla)

This has large roundish leaves with eight corners, which are finely serrated at the edges, and when they first spring up, they are plaited or folded together. The stalks are small and weak, bearing clusters of yellowish green flowers. It grows in gardens and sometimes in meadows and woods, and flowers in May. This has the same virtues as Sanicle, being a species of it.

It removes the pain and heat of all inflamed wounds, ulcers and boils. It can be applied to women's breasts to make them small and firm. It stops internal bleeding and menstrual flow. It can cure burns and scalds, and it is also good for encouraging conception in women.

LARKSPUR
*(Delphnium vel consolida
regalis)*

The dark green leaves are round and tender and divided into many deep sections. It has a straight round stem about 3 ft high; the flowers grow on the tops of the branches in long spikes, sometimes blue, sometimes white and sometimes of a carnation colour. It grows in gardens and also wild, flowering for most of the summer.

If the juice is sniffed into the nose it is good for headaches and migraine. It also helps to heal wounds.

**COMMON
LAVENDER** or
Narrow Leaved
Lavender or Lavender
Spike
(Lavendula augustifolia)

It has a hot dry nature. It provokes urination and menstruation and expels the afterbirth, and also a still-born child. The flowers cure palpitations of the heart, and jaundice and are particularly good for apoplexy, vertigo and palsy.

LAVENDER COTTON
(Abrotonum foemina, Chamaecyparissus)

It is planted in gardens, flowering in July and August. This has much the same virtues as Sothernwood.

It provokes urination and menstruation and breaks up stone. It is also beneficial for the liver, chest and womb. It clears obstructions and dissolves swellings. Taken inwardly, it is a specific medicine against worms.

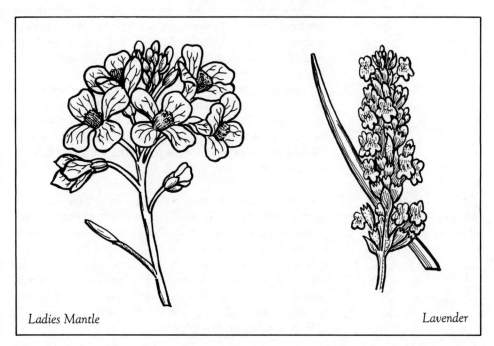

Ladies Mantle

Lavender

LEADWORT or Toothwort
(Dentario cue dentillaria)

The leaves are long, narrow and of a whitish-green colour; the stalks are slender, and the flowers are purple and grow in short thick spikes.

It has a hot biting nature. If masticated, it is good for a toothache. (In fact, it is such a good remedy that, it is said, it will cause a cure even if it is merely held in the hand.) A decoction made with Plantain, wild Tansy, Slow Tree bark, allum and honey of Roses, will fasten loose teeth. A decoction of it alone will provoke urine and break up stone.

GARDEN LEEK
(Porrum)

It is effective in cleansing the chest and a bath made of it with sea water opens obstructions of the womb and provokes menstruation. It will stop bleeding, especially of the nose and mixed in a powder with Frankincense and applied as a poultice, will dissolve swellings. It is exceeding good to make women fruitful.

LEMON TREE
(Limonia malus)

Lemon-trees are preserved from the inclemency of the air in this country by several curious gentlemen, who keep them in greenhouses.

The leaves of this tree have a very fragrant smell and the fruit quenches a thirst, revives the appetite and is useful in fevers of all kinds, being cooling. The juice, mixed with the salt of Wormwood, stops vomiting and strengthens the stomach.

However, I do not take lemon to be good for tender constitutions, for its bitter little particles grate too much on the fibres of the stomach, and induce belchings, flatulence and headaches. Moreover, there is another reason why it cannot be wholesome, for the fruit, when it is sent to us, is not fully ripe.

LENTILS or
Common Lentils
(Lens vulgaris)

The leaves are small, grow on short stalks, and have clasping tendrils by which they stick fast to anything that touches them, and the flowers are small and white. It grows in fields, flowering in May, with the seed ripening in July.

The first decoction aids digestion, and the second decoction is effective in stopping diarrhoea and excessive menstrual flow. A meal of lentils mixed with honey, cleanses ulcers and rotten sores; boiled in vinegar, it dissolves lumps and swellings.

LETTUCE
(Lactuca)

It creates a good appetite, quenches the thirst, aids digestion, promotes sound sleep, provokes urine and increases milk in nurses. The green leaves when bruised, can be applied to burns, scalds, boils and inflammations.

LICORICE
(Glycyrisa liquiritia)

It is planted by us in gardens, and it is brought to great perfection in my Lord Kingstons Gardens at Mitchelstown, where the root may be seen as thick as a man's finger. It is extraordinary pleasant to the taste, far exceeding what is commonly brought to us from foreign countries, so you see, how by a little industry, the most exotic plants may be brought to perfection in this country, which demonstrates what a fertile, prolific land we live in.

The virtues of this plant are so well known, almost to

everyone that I need not enlarge much upon it. The roots are beneficial for coughs and obstructions of the lungs; they also provoke urine, break up stone and are good for the liver and disorders of the mouth.

WHITE LILY
(Lilium album)

It grows in gardens, flowering in June.

The roots are useful in softening and easing pain in swellings and cysts. They are often taken inwardly as an emetic.

LILY OF THE VALLEY
(Lilium convallium)

By distilling the flowers in strong wine, and drinking a spoonful, speech will be restored after an apoplexy; this remedy is also effective for gout and paralysis. A small amount of the root, crushed and taken in wine and vinegar is a great antidote to poison, as it causes the patient to perspire, and is also useful in an apoplexy.

Take 2 drams of the flowers with a dram of the root of White Hellebore and let them be mixed and pulverized, and then sniff the powder up the nostrils. This will be found to be a singular remedy for apoplexy.

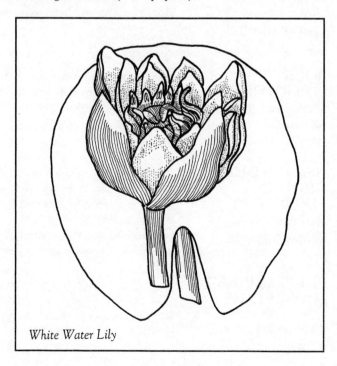

White Water Lily

WHITE WATER LILY
(Nymphaea alba major aquatica)

It grows in pools, rivers and large lakes, flowering in June.

It has a cold dry nature, lacking in sharpness. A decoction of the roots or seed is very effective against sexual desire or fleshly lust. A preparation of the flowers is useful against fevers and headaches, for it causes sweat and quiet sleep, and it also helps to cure diarrhoea, dysentery, gonorrhoea and flows of any kind. The roots applied to wounds, prevent bleeding and the leaves heal boils and inflammations.

YELLOW WATER LILY
(Nymphaea lutea)

This kind of water lily bears yellow flowers; the leaves are large and round and lie on the surface of the water and they are found in pools, rivers and large lakes.

It has the same virtues of the former, but in a weaker degree. It is cooling and binding, and is effective against boils and inflammations.

COMMON LIVERWORT or
Ground Liverwort
(Hepatica vulgaris, Lichen, Hepatica terrestris)

It grows in moist shady places.

It has a cold dry nature, and is very good against inflammations and obstructions of the liver and blood, fever, gonorrhoea and jaundice. If applied externally, it stops bleeding and heals itchy and spreading scabby skin.

ROCK LIVERWORT
(Lichen seu hepatica paetraea)

It grows on rocks and in stony places near rivers, where they are shaded from the sun. It has the same qualities and virtues as the former.

ASH-COLOURED GROUND LIVERWORT
(Lichen terrestris cinereus terrestris)

It consists only of thick crumpled leaves, ash-coloured on the upper side and somewhat whiter underneath, and it bears no flowers or seeds. It grows all year round in dry barren places.

If crushed and taken internally, it is effective against the bites of mad dogs and other venomous animals.

PURPLE SPIKED LOOSE STRIFE or
Spiked Willow Herb
(Lysimachia purpurea spicata)

It grows in moist ditches, flowering most of the summer.

The distilled water is good for sore eyes. A salve made from the leaves is useful in preventing inflammations in wounds.

YELLOW LOOSE STRIFE or Yellow Willow Herb
(*Lysimachia lutea*)

It grows in moist places, and by riversides.

It is very good in stopping all flows of blood, such as nosebleed and excessive menstrual flow and also in sealing wounds.

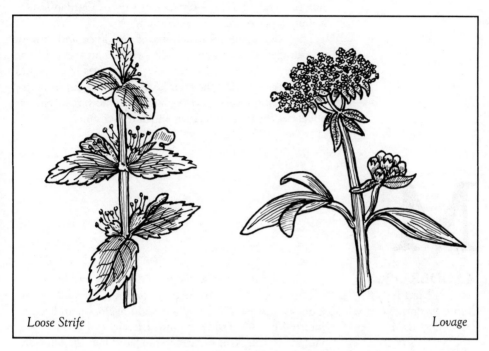

Loose Strife

Lovage

LOVAGE
(*Levisticum*)

It grows in gardens, flowering in June and July, and has a hot dry nature.

It expels flatulence from the stomach and aids digestion, provokes urination and menstruation, clears the sight, and removes spots, freckles and redness from the face.

TREE LUNGWORT
(*Muscus pulmonarius lichen arborum*)

It is a kind of moss which grows on trees and somewhat resembles liverwort only that it is larger, having great scales placed one upon another. It is green on the outside and ashen coloured underneath. It has a bitter taste and a binding nature.

It is very good for inflammations and ulcers of the lungs, if it is crushed and drunk in water. Drinking a decoction of it in wine stops spitting and pissing of blood, diarrhoea and excessive menstrual flow. It is very good against yellow jaundice.

LUPIN or White
Garden Lupin
(*Lupinus sativus alba*)

It has round hard stalks on which grow many leaves, and the flowers grow in white spikes on the tops of the branches, shaped somewhat like Peas Blossoms. It flowers in June. There are also Wild Lupins, which bear yellow and blue flowers. These have the same virtues of the Garden Lupin, but are more powerful in operation.

The seed opens obstructions of the liver and spleen, encourages menstruation, kills worms and expels the stillborn child. If used externally, it removes freckles and spots, and beautifies the face. It can be used against lumps, abscesses and ulcers. A decoction of the flowers in vinegar alleviates the pain of sciatica and cures scrofula.

M

MADDER or Red
Garden Madder
(*Rubia tinctorum,
Sativa hortensis*)

It is planted in gardens, and flowers in May and June.

The roots open obstructions of the liver and spleen, kidneys and womb. They are good against jaundice and internal bruises and, if pounded finely and applied with vinegar to the skin, it cures itching and mange. They also provoke urine, but care must be taken, for they may make the patient piss blood.

**WILD HEDGE
MADDER &
LITTLE FIELD
MADDER**
(*Rubia sylvestris,
Rubeola arvensis*)

These are like the garden madder, but smaller and less rough; the roots are very small, tender and sometimes reddish, and the flowers are not so white.

They are not made much use of in medicine, but they are useful in opening obstructions of the liver and spleen, and they also cure jaundice.

**TRUE
MAIDENHAIR**
(*Adjanthum vulgare seu
capillus veneris*)

The leaves are small, round and serrated, and the stalks are black, shiny and slender, nearly 1 ft high. It grows on stone walls and rocks.

It helps cure asthma, coughs and shortness of breath. It is good against jaundice, diarrhoea, spitting of blood and the bites of mad dogs. It also provokes urination and

Madder

menstruation, and breaks up stone in the bladder, spleen and kidneys.

BLACK MAIDENHAIR
(Trichomanes mas sive polytrichum)

It grows about 3 in high. The stalks are smooth, slender and black, the leaves are deeply indented, sharp pointed and shining green. The back of the leaves have the edges covered with a brown dusty seed. It grows in shady places and at the roots of trees.

Its virtues and qualities are similar to those of the former.

WHITE MAIDENHAIR or Wall Rue or Tent Wort
(Adjanthum album seu ruta muraria)

The leaves are set on short stems, resembling the leaves of Garden Rue, but they are smaller and somewhat indented. The stalks are slender and whitish and it seldoms grows higher than 3 in. It grows on old walls and ruinous buildings.

It has the same qualities as the other Maidenhairs. Taken internally it provokes urine, and breaks up stone in the bladder and the kidneys. The juice is effective in removing films and cataracts from the eyes.

GREAT GOLDEN MAIDENHAIR or Goldilocks
(Adjanthum aureum vel polytrichum aureum majus)

This is a large kind of moss, about 4 or 5 in high, bearing hard, short leaves, with the seed vessel covered with a woolly reddish yellow cap. It grows on healthy and boggy ground.

This has much the same virtues with the former Maidenhairs. Washing the head with a decoction increases hair growth, and prevents hair loss.

SMALL GOLDEN MAIDENHAIR or Little Goldilocks
(Adjanthum aureum minus ceu polytrichum aureum medium)

It grows in boggy places, and while similar in properties to the other Maidenhairs, it is also beneficial for scrofulous conditions.

MALLOWS or Common or Ordinary Mallows
(Malva vulgaris)

It grows in gardens and is very often found growing wild by waysides, and it flowers in May and June.

It provokes urine, breaks up stone and urinary crystals in the bladder and kidneys, aids digestion and cures colic pains. If applied externally, it softens and dissolves all kinds of abscesses and swellings, draws out splinters and eases wasp and bee stings.

MARSH MALLOWS
(Althea, Bismalva, Ibisius)

It is usually planted in gardens, but it sometimes grows in salt marshes, and flowers in July.

If a decoction of the root is drunk in wine, it is very good against urinary crystals, dysentery, coughing and hoarseness. When applied externally, it heals wounds, and dissolves, softens and suppurates cysts, abscesses and swellings. The leaves applied with oil heal burns and scalds and also bites of mad dogs, and wasp and bee stings. An ointment made from it is good for stitches and pains.

VERVAIN MALLOWS
(Alcea vulgaris major, Maloa sylvestris)

The stalks are round, straight and hairy and the flowers have a clear red or carnation colour. The cheese-like seed vessel is larger and blacker than that of the Common Mallows. It grows in hedges and along the borders of fields, and flowers in June.

If a decoction of the root is drunk in water or wine, it stops dysentery and heals internal wounds and ruptures.

MAPLE TREE
(Acer majus)

The timber of this tree is hard, the leaves are broad and five cornered, very much like the leaves of Sanicle, and the fruit is long and flat.

The roots, pounded in wine and drunk are beneficial for pains in the side.

MARIGOLD
(Calendula ceu caltha vulgaris)

(Marygolds are called calendulae because they flower at the beginning of the month.)

They provoke menstruation, cure the inflammation of the eyes and trembling of the heart. Because they promote perspiration, they are frequently used to eliminate smallpox and measles. They are also good against jaundice.

CORN MARIGOLD
(Chrysanthemum segetum vel cellis lutea major)

It grows in cornfields, flowering in Summer.

It has a hot dry nature. A decoction drunk in wine, cures jaundice, and the seed has the same virtues.

MARSH MARIGOLD
(Caltha palustris ceu populago)

It grows in marshy grounds, and has qualities similar to the former.

SWEET MARJORAM
(Majorana, Sampiucus, Amaracus)

It is planted in gardens, and flowers in July and August.

It has a hot dry nature. It provokes urination and menstruation, opens obstructions of the liver and spleen, is good for colic pains and for disorders of the head and nerves, such as apoplexy, epilepsy and migraine. A plaster made of it with oil and wax, softens growths and is good to be applied to dislocations.

WILD MARJORAM
(Origanum vulgare)

The stalks are brown, hairy and brittle, and about 1 ft high; the leaves are broad, round-pointed and brownish-green, and the flowers are small and purple. It grows in hedges and thickets and flowers in July.

Wild Marjoram has a hot dry nature. It is good against pains of the stomach and heart and also useful for coughs, pleurisy and obstructions of the lungs and womb, and it also comforts the head and nerves. Externally applied, it is beneficial for swellings of the ears, ulcers of the mouth, dislocations, bruises and eruptions of the skin. The distilled oil eases toothache.

MASTERWORT or False Pellitory of Spain
(Imperatoria seu astrantia)

It bears large rounded leaves, serrrated about the edges, among which grow tender knotty stalks, about 1 ft high, and on the tops of which are clusters of small white five-leaved flowers. It is cultivated in gardens, and flowers in July.

It has a very hot and dry nature. The root is a good antidote to poison and infection, cures old fevers, induces perspiration, helps digestion, quickly heals internal bruising, bites of mad dogs, bowel disorders, and all nervous conditions of the head. Made into a poultice, it cures abscesses and sores.

HERB MASTICK
(Marum vulgare)

It is planted in gardens, flowering in June and July.

Taken internally, a decoction in wine is useful against cramps and ruptures, and for easy urination. It encourages menstruation and expels a stillborn child, and the after-birth. It is also good for disorders of the head and nerves. This has much the same virtues as sweet Marjoram.

MEADOW-SWEET or Queen of the Meadow
(Ulmaria, Regina prati)

It grows in moist meadows and by riversides, flowering in June.

A decoction or powder made from the roots stops diarrhoea, dysentery and all kinds of flow, while the flowers infused in white wine are beneficial for fevers.

MEDLAR TREE
(Mespilus sative)

It grows only in gardens and it flowers in May, with the fruit ripening in November.

Medlars have a cold, dry astringent nature. When hard and green, they are useful in stopping diarrhoea. If the crushed stones of the Medlar are drunk in a solution, they break up stone in the bladder.

COMMON MELILOT or Garden Melilot
(Melilotus vulgaris)

It is planted in gardens, flowering in June.

It is good against boils and all kinds of swellings. The juice dropped into the ears eases their pain, and applied to the forehead with oil of roses and vinegar, cures a headache, and dropped into the eyes, clears the sight, and dissolves

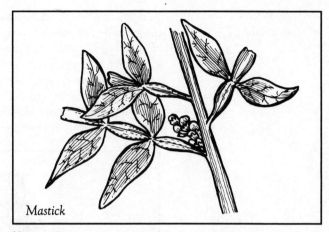

Mastick

films and cataracts. If the herb is mixed in a plaster, it will draw out infection from wounds. A decoction in wine provokes urine and expels urinary crystals.

WILD MELILOT
(Melilotus sylvestris)

It frequently grows in cornfields, upon high banks and in hedges, and it flowers in June.

This Melilot is stronger and more effectual in its operation than the former. It dissolves abscesses, sores and swellings, cures jaundice and disperses hard obstructions. A decoction is good against palpitations of the heart.

The effectiveness of this herb was made known to me by a Gentlewoman of my acquaintance who had a swelling for a year or more on her right side, which was cured by three or four times rubbing the grieved part with an oil made of this herb, when as before all other medicines proved ineffectual.

COMMON MELON
or Musk Melon
(Melo vulgaris)

Sown yearly in the Spring, they ripen in July and August.

These have a cold moist nature. The seed is good against fevers, and help provoke urine and break up stone. Melons must be eaten moderately.

ENGLISH MERCURY or Good Henry or All Good
(Bonus henricus, Tota bona, Maercurialis, Lapathum unctuosum)

The leaves grow on long stalks and are of triangular shape (like Spinage) and are yellow–green in colour; on the tops of the stalks are spokes of small herbaceous flowers. It grows in waste places and among rubbish, and flowers in Spring.

Drinking a decoction aids digestion and purges cold phlegm. The young shoots are good for scurvy and provoke urine. Pounded with butter, it can be used as an enema.

MILKWORT
(Polygala vulgaris)

This is a small herb with slender pliant stalks creeping on the ground. The leaves are small and narrow, like the leaves of small Hyssop, and the flowers, which are of a blue colour and which are very much like the flowers of Fumitory, are succeeded by small pods, like those of Shepherd's Purse. It grows on dry healthy ground, and flowers in May.

It has a hot moist nature, and it encourages the production of milk in nursing mothers.

Milkwort

GARDEN MINT or Spearmint
(Mentha sativa)

It is planted in gardens, flowering in July.

It is good for diseases of the mouth and stomach, stops vomiting, gonorrhoea and excessive menstrual flow. it also expels worms from the body.

WATER MINT
(Mentha aquatica seu sisymbrium)

It grows near rivers and in low lying meadows, flowering in June and July.

It is aromatic, and good for most disorders of the stomach, expelling wind from it. It also opens obstructions of the womb and provokes menstrual flow.

WILD MINT or Horse Mint
(Menthastrum seu menthastrum)

Being of a hot nature, it is very good in dissolving swellings and is useful when applied to dislocations. It helps cure flatulence and colic pains.

MONEYWORT or
Herb Two Pence
*(Nummularia major
vulgaris lutea)*

The stalks are small and slender and creeping on the ground, shaped somewhat like a silver penny, and the flowers are yellow like gold cups. It grows in moist meadows near ditches and watercourses, flowering for most of the summer.

It has a dry nature. Drinking a decoction in wine and honey heals wounds and ulcers of the lungs. It is also very good against coughs, especially the chin cough (whooping cough) which children are subject to.

MOON-WORT
(Lunaria)

This plant bears only one leaf, which is cut into several sections standing around the middle of the stalk; on the top it bears several bunches of small globular heads, in which the seed is contained; it seldom grows more than 3 in high. It grows on dry grassy hills or mountains, appearing only in May and June, but disappears afterwards.

It is good for healing all kinds of wounds that are curable and also to stop bleeding. If an ointment made from it is applied to the kidneys, it can cure dysentery.

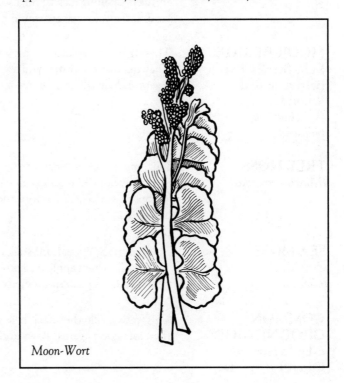

Moon-Wort

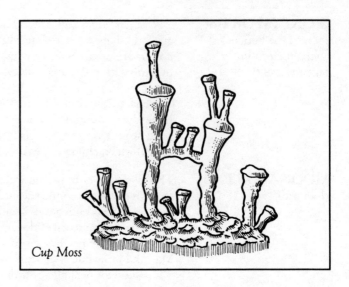

Cup Moss

The common people say that this herb has a singular virtue of opening locks, and also, that if a horse were only to trod upon it, his fetters would presently drop off.

MOOR BERRIES or Moss Berries, Bog Berries, or Red Whortes
(*Vaccinia rubra palustria*)

These berries are usually preserved and made into tarts. They quench the thirst, and are good against hot and acute fevers and all inflammations of the blood.

TREE MOSS
(*Muscus arboreus*)

This moss has a dry astringent quality. A decoction externally applied is good against excessive menstrual flow. It stops nosebleeds if put into the nose and also other flows of blood.

SEA MOSS or Coralline
(*Corallina*)

This has many small hard stalks of a stony substance.
It is useful when applied to boils and gout. Crushed into a solution, it is a powerful remedy for worms.

COMMON GROUND MOSS
(*Muscus terrestris vulgaris*)

This moss like the rest, has a binding quality, and is therefore good against flows and haemorrhages of all sorts.

MOSS (Growing on Dead Men's Skulls) *(Muscus innatus cranio humano seu usnea)*

This has a binding quality, is good for stopping nosebleeds and other haemorrhages.

CUP MOSS *(Muscus pixidatus)*

This moss has many hoary or whitish green leaves, spread on the surface of the ground, among which grow whitish dusty cups, a $\frac{1}{4}$ in high. It grows on banks and dry barren ground.

If a decoction sweetened with sugar is drunk, whooping cough will be cured.

MOTHER OF THYME or Wild Thyme *(Serpyllum vulgare)*

The stalks are slender and woody; the leaves are like the leaves of Garden Thyme, only larger. The reddish purple flowers grow on the tops of the stalks, whorl fashion. It grows in untilled stony places, flowering in June and July.

It is good for diarrhoea, spitting or vomiting of blood, cramps, catarrh and old coughs, it aids digestion and lack of appetite. It provokes menstruation and is useful for all disorders of the head and nerves, if a decoction is taken internally. The juice externally applied with oil of roses and vinegar eases headache, and the distilled oil eases a toothache.

Mother of Thyme

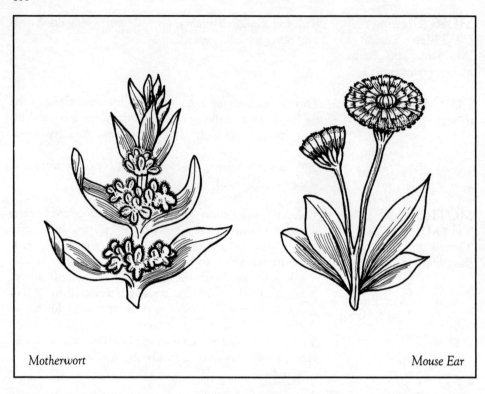

Motherwort

Mouse Ear

MOTHERWORT
(Cardiaca)

It has square brown woody stalks; the leaves are large, broad, deeply indented and have three sharp points, like Nettle or Horehound leaves, and the flowers have a reddish purple colour. It grows in untilled places, around old walls and lanes, flowering in June and July.

If bruised and applied to wounds it prevents inflammation and abscess, and stops them bleeding and heals them. It is good against disorders of the heart, such as palpitations and it is also used against disorders of the spleen and womb. Birth can be facilitated by drinking a dram of crushed Motherwort in wine.

MOUSE EAR
(Auricula muris vel pilosell)

It grows under hedges and along the borders of fields, flowering in June and July.

It has a dry nature, and is good against the spitting of blood, all kinds of flow, coughs, ulcers of the lungs, mouth and eyes, and shingles.

MUGWORT or the Great Common Mugwort
(Artemisia major vulgaris)

It grows along the borders of fields, in highways and waste places, flowering in June.

If pounded and applied with the oil of sweet almonds, it cures the pain of the stomach. If the juice is mixed with the oil of roses and applied, it will ease pains or aches in the joints and cramps. It also provokes menstruation. If it is added to barrels of ale, it will prevent them from souring.

MULBERRY TREE or Common Black Mulberry-tree
(Morus)

This tree grows in gardens, and the fruit ripens in August and September.

The unripe mulberries are cold, dry and astringent. They stop diarrhoea, dysentery and vomiting. If applied with oil, the leaves heal burns and scalds. The bark of the root of the tree is cleansing and abstersive. It opens obstructions of the liver and spleen, aids the digestion, expels worms from the body, and if held in the mouth, eases a toothache. The root, if cut about the latter end of harvest, will yield a gum which is also very good for toothache. The ripe fruit is cooling and good for alleviating the heat of burning fevers and creating an appetite.

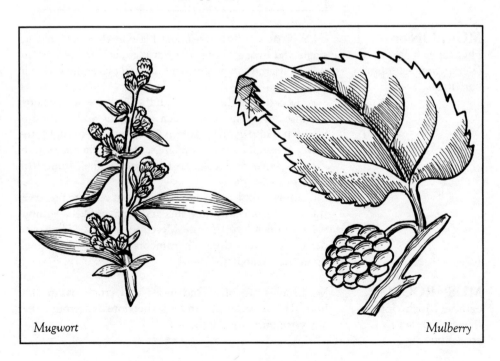

Mugwort

Mulberry

Mullein

MULLEIN or
Hightaper
*(Verbascum vel tapsus
barbatus)*

It has broad woolly leaves about 1 ft long; the stalk is round, single and hoary, about 5 or 6 ft high, and the flowers are yellow. It grows in gardens and in wild places, flowering in July.

A decoction of the leaves with the flowers is good against the diseases of the lungs, such as coughs and spitting of blood. It is also good for colic pains. Applied externally, the leaves are useful against swellings, ulcers, inflammations of the eyes, burns and scalds. A fomentation made from them is a remedy for piles.

Country people say that carrying it about one preserves the wearer from enchantments and witchcraft, not without some justification, because Mercury gave it to Ulysses to defend him from the enchantments of Circe – but I look upon this as fabulous.

MUSHROOM or
eatable Mushroom
*(Fungus campestris
esculentus)*

Mushrooms are of a cold moist and crude nature and hurtful to the stomach, and it is therefore dangerous to eat any great quantities of them.

GARDEN MUSTARD
(Sinapi)

It grows in waste places, amongst rubbish, and is frequently sown in gardens, flowering in June.

Mustard seed has a very hot and dry nature. It aids the digestion, warms the stomach, creates an appetite, loosens phlegm, provokes urination and menstruation, and is good against scurvy, apoplexy, lethargy and paralysis, especially of the tongue. A gargle made of it with honey and vinegar is useful against tumours of the uvula and the glands of the throat. This would also be useful against fevers. It is exceeding good for any one that is troubled with gout for, if the seed is pounded and applied in a plaster, it will raise a blister which will carry off the malignity. I have been acquainted with some gentlemen who, by this means, got rid of it for several years.

TREACLE MUSTARD or Penny Cress
(Thlapsi)

It has a very hot and dry nature. The seed encourages menstruation, breaks up internal abscesses, and purges bilious matter. It also provokes urine, and is useful against gout and sciatica.

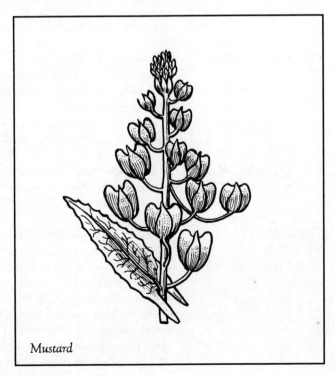

Mustard

MYRTLE-TREE
(Myrthus)

The leaves and berries have binding properties and are good against haemorrhages. They are also generally effective against disorders of the lungs and against running eyes and noses.

N

NAVELWORT or
Wallpennywort
(Cotyledon vel umbilicus veneris)

The leaves are round, thick and succulent; the stalks are about a span long, on the tops of which grow the flowers in long spikes of a whitish or carnation colour. It grows on old stone walls, and flowers in May.

It has a cold moist nature. The juice is a good remedy against all inflammations, chilblains, burning fevers and piles. Taken internally it provokes urine, and cures disorders of the liver.

SWEET NAVEW
(Napus dulcis)

The leaves are large, like the leaves of Turnip, but are smoother, while the stalk grows two or three feet high and on which grow many yellow flowers. It grows in gardens, and flowers in April.

The seed is good against poison as well as small-pox and measles. It also provokes urination and menstruation.

WILD NAVEW
(Napus sylvestris)

This is smaller than the Sweet Navew and its leaves are more jagged, but otherwise it is similar both in appearance and properties. However the seed is hotter and stronger.

STINGING NETTLE
(Urtica major vulgaris urens)

It has a hot dry nature. The seed mixed with honey and taken internally, cleanses the lungs of phlegm, it is also good for shortness of breath and whooping cough. A gargle of the juice of the leaves is useful for a fallen womb and about half a pint of this juice taken internally stops vomiting and spitting of blood, and all haemorrhages and flows. If crushed leaves are applied to wounds, they stop the bleeding or if pounded with salt, they are useful for the bites of mad dogs. The root provokes urine and is a remedy for jaundice.

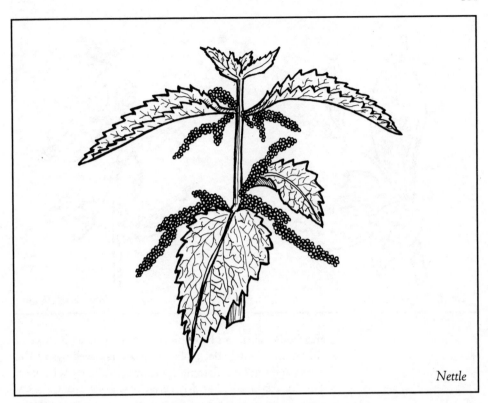

Nettle

COMMON GARDEN NIGHTSHADE
(Solanum hortense vulgare)

It has round, brittle, succulent stalks, the leaves are large, blackish, soft and full of juice, the white flowers grow in clusters and it bears berries of a black shiny colour when ripe. It grows about old walls, under hedges and near highways, and flowers in August, with the berries ripening in September.

It has a very cooling astringent nature. It is a good cure for burning fevers, boils, ulcers, inflammations, burns and scalds.

DEADLY NIGHTSHADE or Dwale
(Solanum lethale)

It has tall blackish angular stalks about 5 or 6 ft high, on which grow broad dull green leaves; the brown bell-fashioned flowers are succeeded by black shining berries as large as cherries. It is found in woods and hedges, with the berries ripening in August.

The leaves and fruit have a very cold nature. The green

Deadly Nightshade   *Woody Nightshade*

and fresh leaves can be applied externally as with those of Common Nightshade, to burning fevers, swellings of the breast and similar inflammations, but caution must be used. The fruit is deadly, and if eaten, it causes a deep sleep, followed by an outrageous passion of anger which continues until death.

WOODY NIGHTSHADE or
Bitter Sweet
(Solanum lignosum seu amara dulcis)

It has small tender branches of a woody substance, the leaves are smooth, green and sharp-pointed, like the leaves of Ivy, and the purple flowers consist of one leaf, which is laid open like a star. The berries, when ripe, are of a pure shining red. It grows in moist places, about ditches and ponds, and flowers in May.

It has a hot dry nature. A decoction in wine taken internally opens obstructions of the liver and spleen, and is therefore good for jaundice. It also heals all internal wounds, bruises and ruptures, for it dissolves congealed blood, causing it to be passed by the urine.

NIPPLEWORT or
Tetter-Wort
(Lampsana)

This herb is good for sore nipples, and helps ease painful eruptions on the skin.

O

COMMON OAK TREE
(Quercus vulgaris)

All parts of the oak have a binding nature, and are therefore useful against diarrhoea, dysentery, haemorrhages and flows of all kinds. The bark can be used in gargles for dropped uvula.

There is also a fungus which grows on this tree (Incrementum fungosum querceum); this too can be used in medicine.

OATS
(Avena)

Oatmeal has a hot, dry and astringent nature, not easily digested. Therefore it is fit only for labourers who can expel its heat by perspiring freely, otherwise it is apt to cause itchiness. Hot oats put into a bag and applied to the grieved spot are good for pleurisy and bowel disorders.

Oak

ONION
(Cepa)

Onions have a hot nature, creating wind in the stomach. They provoke urine, aid digestion and are good for coughs and diseases of the breast. Being externally applied with oil and vinegar, they cure piles. The juice dropped in the eyes, cleanses them, and if dropped into the ears, it is good against deafness and a humming or ringing in the ears. Sniffed into the nose it causes sneezing, and it therefore clears the brain. Mixed with salad oil, it is exceeding good to be applied to burns and scalds before they rise to blisters. If mixed with rue and salt, it can effectively be applied to dog bites. It will also restore hair to parts that are bald.

BOG ONION
(Cepa palustris)

This grows in bogs. The root is bulbous and is divided into cloves like Garlic.

A clove of this placed in a glass of water and left for half an hour, will make the water very thick and ropy, and it can then be beneficially applied upon a cloth to dislocations.

COMMON WILD ORAGE or Lambs Quarter
(Atriplex sylvestris vulgatior)

It is cooling and moistening, aiding digestion. The leaves are frequently boiled and eaten with salt meat, like Colewort. The seed opens obstructions of the liver and is therefore good against jaundice, and when crushed, it is effective against boils, inflammations, burning fevers and gout.

STINKING ORAGE or Arrach
(Atriplex olida)

It grows in rich soil, on dunghills and near highways.

It provokes menstrual flow, facilitates labour and expels the afterbirth, if a decoction is taken internally.

ORANGE TREE
(Aurantia malus)

Orange trees grow plentifully in foreign countries, but of late years they have been transplanted here, and now, by the industry and cultivation of curious gentlemen are, in some gardens brought to perfection. I have seen about seventy or eighty oranges taken off one tree in the Right Honourable, the Lord Kingstons Garden at Mitchelstown, and these were as good as any I have seen brought hither from Spain, or the West Indies, so you see what a prolific and fertile soil we live in, where the most exotic plants might, by a little care and industry, flourish.

Oranges are good against scurvy and diseases of the mouth, and prevent nausea and vomiting. They are an effective antidote to infection and the seed kills and expels worms.

Orpine

Ox Eye

ORPINE or Live Long
(Crassula vel fabaria)

It has round brittle stalks, on which grow thick, fat, oval leaves indented about the edges, and the flowers grow in small purple clusters. It is found in gardens, but grows wild in moist shady places, and flowers in June.

It is binding and cooling, good against dysentery. It can be used externally against boils, inflammations, scalds and burns.

OSIER
(Salix aquatica folio longissimo)

It grows in moist places and by riversides.

It has a cold dry astringent nature. A decoction of the leaves and bark in claret taken internally, is good against spitting or vomiting of blood, excessive menstrual flows and all other flows. The ashes of the bark, mixed with vinegar and applied to warts causes them to fall off. It also takes away callous or hard skin on hands and feet and the sap is good for inflammations of the eyes and for blood-shot eyes.

OX-EYE
(Buphthalmus)

It is about half a cubit high, with three or four stalks bearing tender winged leaves, like Yarrow, and the flowers have a bright yellow colour, almost like the flowers of Marygolds. It is planted in gardens and also grows wild, and it flowers in June and July.

It has a hot dry nature. The flowers crushed and mixed with oil and wax are beneficial when applied to swellings. A decoction in wine is good for jaundice.

P

PALMI CHRISTI or the Greater Spurge
(Cataputia major vel ricinus)

It has a smooth, round stalk taller than a man, and the leaves are large and roundish, cut into five, seven or nine sections and serrated about the edges. The flowers grow in clusters like grapes, and are succeeded by triangular husks, which contain white seeds, which are somewhat smaller than kidney beans.

The seed crushed and taken in whey purges phlegm and bile. It is also good against jaundice. The oil extracted from the seed is beneficial against itching and kills lice in children's heads.

GARDEN PARSLEY
(Petroselinum vulgare seu apium hortense)

This aids both appetite and digestion. A decoction of the root and seeds taken internally open obstructions of the liver, kidneys and all internal organs. It provokes urine, and breaks up and expels bladder stone. It is also good for jaundice.

MACEDONIAN PARSLEY
(Petroselinum macedonicum)

It grows in gardens and is of a hot and dry nature.

The seed provokes urination and menstruation, breaks up stone in the bladder and kidneys, and expels flatulence from the stomach.

WILD PARSLEY
(Apium sylvestre)

It bears large jagged leaves like those of Wild Carrot, but larger. The stalks are round and hollow, 4 or 5 ft long, of a brown red colour, on the tops of which grow round tufts of white flowers. It is found in moist places about ponds and ditches, flowering in June.

It has a hot dry nature, and if the root is held in the mouth and chewed, it eases a toothache.

PARSNIP
(Pastinaca sativa)

Parsnips have a hot dry nature, and they provide better nourishment than Carrots. They provoke urine, and are beneficial for the chest and kidneys.

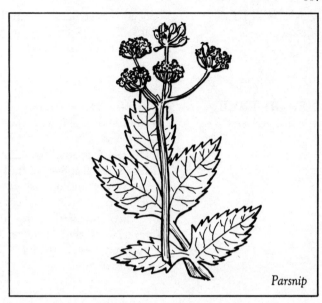

Parsnip

WATER PARSNIP
(Sium latifolium)

It grows in rivers and moist ditches, and flowers in May and June.

The leaves provoke urine, expel urinary crystals, and remove obstructions of the liver, spleen and kidneys. Externally applied, they are good against cancerous tumours in the breast.

COW PARSNIP
(Sphondilium)

The stalks are long, round and full of joints, the leaves are rough and dark green, and on the tops of the stalks are clusters of white flowers. It grows along the borders of fields and meadows, and flowers in July.

It has a hot nature, and when bruised and applied to swellings, it disperses and dissolves them.

WILD PARSNIP
(Pastinaca sylvestris)

The leaves, flower and seed are much like those of Garden Parsnip, although the leaves are smaller, the stalks more slender, and the root harder, smaller and not good for eating. It grows under hedges and near highways and paths, flowering in June.

The seed is very good against all poison, and the bites of venomous creatures. The roots and seed provoke urination and menstruation, encourage flatulence and are beneficial for the liver and spleen.

GARDEN PEA
(Pisum)

Peas encourage flatulence in the body, but not as much as beans. When green they are a pleasant nourishing food, which sweetens the blood.

PEACH TREE
(Persica malus)

It is planted in gardens, flowering in March and the beginning of April, with the fruit ripening in August and September.

The leaves open obstructions of the liver and spleen and aid digestion. If applied to the navels of young children, they expel worms, and if crushed and applied to wounds, they cure and heal them. The kernels are beneficial for the liver and chest and if they are finely crushed and boiled in vinegar until they dissolve and become like pap, they wonderfully restore the hair. The flowers are purgative and open obstructions.

PEAR TREE
(Pyrus)

Pears nourish more than apples, especially those that are sweet-sour. They are however, very unwholesome if eaten raw, unless wine is drunk after them.

Pears stop the diarrhoea, and can also be applied to fresh green wounds, which they heal.

PELLITORY OF THE WALL
(Parietaria vel helxine)

It grows on old stone walls, and flowers in May.

It has a cold moist nature, and is very useful against bilious inflammations, burning fevers, boils, ulcers, spreading sores and scabs, if crushed and applied with a poultice. It is good for the chest, and breaks up stone in the bladder and kidneys. A decoction taken internally, provokes urine, or a hot poultice may be applied to the region of the bladder.

PELLITORY OF SPAIN
(Pyrethrum)

It is planted in gardens, and flowers in July.

The roots have a very hot nature and are good against apoplexy, epilepsy and toothache. If chewed or held in the mouth, they clear the head. An oil made from them is very good when applied to bruised parts or limbs that have become numbed by the cold or afflicted with palsy.

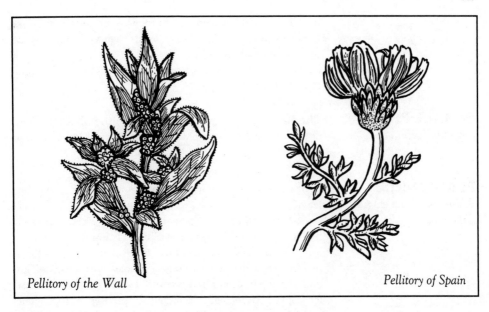

Pellitory of the Wall   *Pellitory of Spain*

PENNY ROYAL
(Pulegium vulgare)

It is cultivated in gardens, and has a hot dry nature.

It provokes urination and menstruation, and it eliminates the stillborn child and the afterbirth, as well as the un-natural birth. It is good for the chest and is useful for opening obstructions of the lungs, and also against coughs and asthmas. If it is boiled in water and vinegar it stops vomiting. A decoction of the leaves sweetened with sugar is a good remedy for whooping cough and also eases the pain of gout.

MARSH PENNY WORT or White Rot
(Cotyledon aquatica)

It has smooth round hollow leaves, somewhat like Wall Pennywort, but the white flowers are smaller, smoother and more deeply indented, and are to be found growing under leaves. It grows in low meadows and moist valleys, flowering in July.

It has no particular value in medicine, but this herb is hurtful to sheep.

MALE PEONY
(Paeonia mas)

It is planted in gardens, flowering in May.

Drunk in honey and water, the powder of the root provokes menstruation and cures the gripes which women are subject to who have just given birth. It opens obstructions of the liver and kidneys. Wearing the root and seed around the neck is useful against convulsions, especially in children.

Sixteen of the black seeds drunk in wine is a specific remedy for nightmares.

FEMALE PEONY
(Paeonia faemina)

This has larger and greener leaves than the Male Peony; its flowers are smaller and paler. It has similar properties to the Male Peony.

PERIWINKLE
(Vinca pervinca clematis daphnoides)

It has many slender, long branches, on which grow oval leaves, much like bay-leaves, only smaller. The flowers are blue, and grow among the leaves in short spikes. It grows in hedges and shady places, flowering in Summer.

It has a dry and astringent nature. A decoction of it taken in wine stops diarrhoea, dysentery, excessive menstrual flow and spitting of blood. If it is bruised and placed in the nose, it stops a nosebleed, if chewed, it eases a toothache. Likewise, it is extraordinarily good for healing bruises.

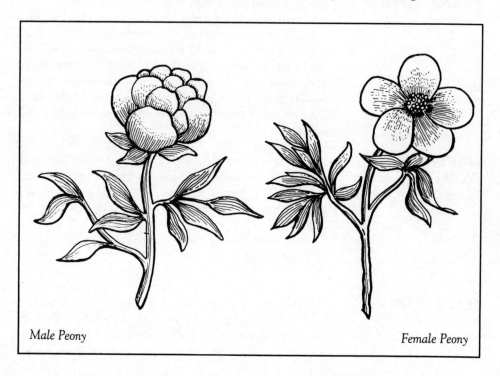

Male Peony

Female Peony

ST PETER'S WORT
or Square St John's Wort
(*Ascyron vulgare*)

The leaves are like the leaves of St John's Wort, only larger and greener, covered with a fine soft down, and having no small holes, while the flowers are paler than those of St John's Wort, but are otherwise similar. It grows in uncultivated places, in hedges and copses, flowering in July and August.

If the seed is drunk in honey and water for a long time, it cures sciatica. Crushed seed is good when applied to inflammations. A decoction of the leaves in wine may be used to wash out and heal wounds.

MALE OR RED FLOWERED PIMPERNEL
(*Anagallia terrestris mas*)

It has small square tender stalks about six inches high. The leaves are small, like the leaves of small Chickweed and full of black specks, and the flowers are a beautiful scarlet colour. It grows in gardens and among potherbs and in cultivated fields, flowering in May and June.

Its properties are similar to those of the Female Pimpernel.

FEMALE PIMPERNEL
(*Anagallis foemina*)

This is like the Male Pimpernel in every respect, except in the colour of its flowers, which are blue.

It has a hot dry nature without any sharpness. Taken

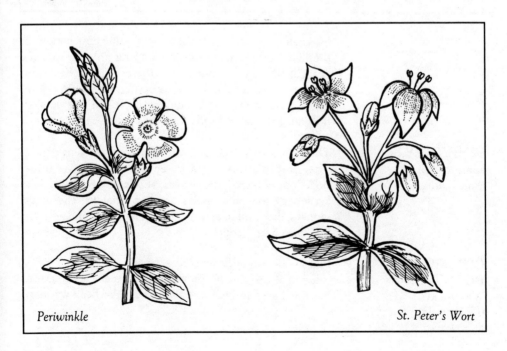

Periwinkle

St. Peter's Wort

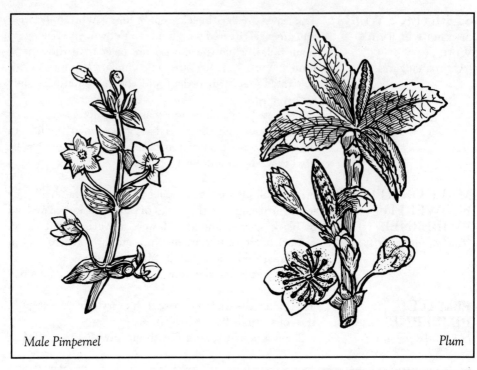

Male Pimpernel   *Plum*

internally, a decoction is good for opening obstructions of the liver and kidneys, and for drawing phlegm from the head, if the juice is sniffed into the nostrils. It heals and cleanses corrupt ulcers and decaying sores, as well as drawing out thorns and splinters. If bruised and applied, they are good against inflammations and hot tumours.

PINE TREE or Manured pine (*Pinus sativa*)

The bark leaves and cones are of a dry astringent nature. They stop diarrhoea and dysentery and provoke urine. Boiled in vinegar, the leaves alleviate toothache. The kernels of the pine apples are beneficial for the lungs, kidneys, liver and spleen. They loosen phlegm and are good for consumptive coughs.

COMMON PLANTAIN or Waybread (*Plantago latifolia vel septinervia*)

It grows by waysides, flowering in May.

It has a hot astringent nature. It is very good against haemorrhages and flows of all kinds; it also heals wounds.

GREAT WATER PLANTAIN
(Plantago aquatica major)

It has broad green leaves and it bears small white flowers, each divided into three parts, which are succeeded by triangular husks. It grows in rivers and brooks, flowering in July.

It has the same properties as Common Plantain.

PLUM TREE
(Prunus hortensis)

Plums are cooling, they alleviate thirst and are somewhat purgative.

POLYPODY or Wall Fern & **POLYPODY OF THE OAK**
(Polypodium vulgare)
(Polypodium quercinum)

It is very good against bowel disorders, and purges bilious substances. If crushed and drunk in honey and water with a little aniseed, it is good for opening obstructions of the spleen, and easing fevers. Made into a powder and sniffed into the nostrils, it is useful against polyps.

POMPKIN
(Pepo)

It has a cold moist nature. If crushed finely, the powder heals inflammations of the eyes when applied to them. If the seed is pounded with Barley Meal and its own juice added, it removes freckles and spots from the face. A dram of the crushed dried root taken in honey and water, excites vomiting.

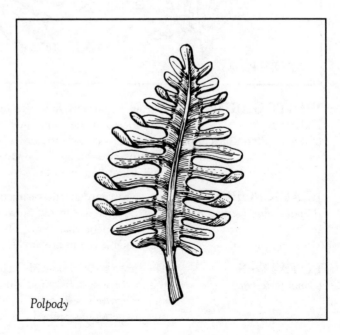

Polpody

WILD RED POPPY
(Papaver rubrum erraticum)

It grows among corn, flowering in June and July.

It has a cooling and refreshing nature. By drinking a decoction of five or six heads in wine, pain is alleviated and sleep is induced. The same result can be achieved by taking the seed with honey. The bruised leaves of the green heads can be applied to boils, hot ulcers and burning fevers. The flowers are useful against inflammatory fevers.

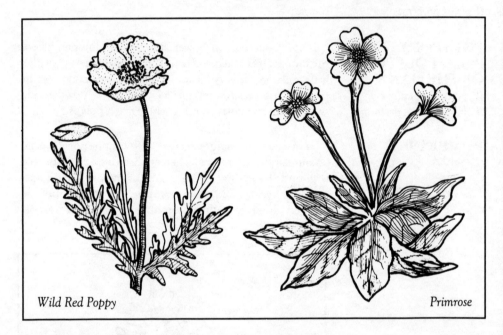

Wild Red Poppy

Primrose

WHITE GARDEN POPPY
(Papaver album sativum)

It grows in gardens, flowering in June, and the heads should be gathered at the end of July.

It eases pain and promotes sleep, and is very good against catarrhs and coughs. The green heads can be applied to boils and inflammations.

BLACK POPPY
(Papaver nigra sativum)

It grows in gardens, flowering in June.

If the seed is drunk in wine it stops diarrhoea, and excessive menstrual flow. The leaves can be applied to inflammations and hot swellings.

POTATOES
(Poma terrestria)

They are a healthy nourishing food (which is evident by the robust constitutions of a vast number of the natives who are almost entirely supported by them). They provoke urine and are generally strengthening.

COMMON PRIMROSE
(Primula veris minor vulgaris)

It grows in thickets and under hedges, flowering in March and April.

It has a dry nature, and is good for disorders of the head and nerves. The juice of the root encourages sneezing, therefore clearing the head of viscous phlegm.

PRIVET or Prim Print or Bindweed
(Lingustrum vulgare)

It has a small shrub, bearing small oblong leaves. The branches are tough and pliant, on the tops of which are thick spikes of white flowers. It grows in hedges, flowering in May and June.

It has a binding and cooling nature, good for ulcers and inflammations of the mouth and throat.

PURSLAIN
(Portulaca)

It grows in gardens and is of a cold moist nature.

It comforts a weak inflamed stomach, and is good against burning fevers, worms, haemorrhages, boils and flows of all kinds. The seed kills and expels worms.

Q

QUICKEN-TREE or Roan tree or Wild Service Tree
(Sorbus torminalis vel mespilus sylvestris non spinosa)

The leaves are cut into seven sharp-pointed segments, pale green above and whitish underneath. It grows very tall and has a whitish bark. The flowers appear in yellowish white clusters, while the fruit is set on long stalks, and is twice as big as the Common Haw. It grows frequently in woods and thickets.

The unripe fruit is cold, dry and astringent. A decoction stops diarrhoea, dysentery and all kinds of flow of blood. When ripe, the fruit is beneficial to the stomach, aiding digestion.

QUINCE TREE
(Malus cydonia vel cotonea)

It flowers in May, with the fruit ripening in September.

Quince stops diarrhoea, dysentery and haemorrhages of all kinds. They also strengthen the stomach, aid digestion and stop vomiting.

R

GARDEN RADISH
(Raphanus hortensis)

It grows in gardens, and flowers in May.

It has a hot dry nature. The finely pounded roots applied with vinegar cure the hardness of the spleen. The seed provokes urination, menstruation and vomiting. Drunk with vinegar and honey, it kills and expels worms from the body.

HORSE RADISH
(Rapahanus sylvestris seu rusticanus)

It has a hot dry nature; and only the root is used. It promotes a good appetite and digestion and is useful against jaundice and scurvy. If made into an ointment, it cures itchy and scabby skin.

WATER RADISH
(Raphanus aquaticus)

It has leaves like the Common Radish, only smaller and more jagged; the stalks are about 1½ ft long, on which grow several yellow flowers, and the root is as thick as a finger. It grows by ditch sides, and near pools and rivers, and flowers in June.

It has a stronger and more biting nature than garden radish and its properties are similar, only more powerful. It is very good in provoking urine.

RAGWORT or St James Wort
(Jacobaea vulgaris)

It grows usually in fields and on banks, and flowers in June and July.

It has a hot dry nature. It is especially good in healing wounds and ulcers; gargling the juice cures inflammations, tumours and abscesses of the throat. A plaster made from it is useful for sciatica. Made into a plaster with oatmeal and butter, it is good against inflammations and all hot tumours.

BROAD LEAVED MARSH or **WATER RAGWORT**
(Jacobaea latifolia palustris)

A poultice of the leaves is accounted good for sciatica, and an ointment made of them clears filthy ulcers.

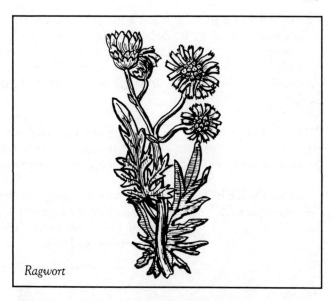

Ragwort

RAMSONS
(Allium ursinum)

It usually has two broad blades, or leaves, almost like the leaves of May Lillies from which grows a stem bearing many small white flowers. It grows in moist shady places, and flowers in April and May.

It has a hot dry nature. Taken internally it kills worms; if bruised and applied to the temples it eases a headache, and applied to the teeth with vinegar, it eases a toothache. It is also good in provoking urination and menstruation, in removing obstructions from the kidneys and chest, and in expelling flatulence.

RASPBERRY or
Hindberry Bush
(Rubus idaeus)

This flowers in May, with the fruit ripening in June and July.

An application of the flowers bruised with honey is beneficial for inflammations of the eyes, burning fever and boils. A decoction is useful for weak stomachs. The fruit is good for the heart and diseases of the mouth.

RED RATTLE or
Lowsewort
(Pedicularis pratensis rubra)

It bears very small leaves deeply indented at the edges, spreading on the ground. The stalks are weak and small, on which appear reddish purple flowers like those of Red Nettle. It grows in moist meadows, and flowers in May.

It has a cold dry astringent nature. Drinking a decoction of it in wine is useful against menstrual flow and all other flows of blood.

YELLOW RATTLE
or Lowsewort or
Coxcomb
*(Pedicularis,
Crista galli,
Lutea)*

It has a straight stem, on which grow long, narrow leaves deeply indented about the edges, while the flowers are like those of Red Rattle, only of a pale yellow colour. It grows in dry meadows, and flowers in June and July.

It has properties similar to the former.

BUR REED or Burr
Flag
(Sparganum ramosum)

It grows in moist meadows, rivers and ditches.

A decoction of burrs is good against the bites of venomous creatures, if it is drunk or the wound washed with it.

COMMON REED
(Arundo vallatoria)

It has a hot dry nature. The finely crushed root draws out thorns and splinters; mixed with vinegar it alleviates the pain of dislocated limbs. The leaves are good against inflammations and boils.

Rest Harrow

REST HARROW or
Cammock or Purple
Rest Harrow
*(Anonis ononis,
Resta bovis)*

It grows along the borders of fields and near roadsides, and flowers in June and July.

It has a hot dry nature. It provokes urine, breaks up stone and urinary crystals, and opens obstructions of the liver and spleen. Externally applied, it cures abscesses and dissolves tumours.

TRUE RHUBARB
(Rhàbarbarum verum)

It has a dry astringent nature. It is good against colic pains, disorders of the stomach, and obstructions of the liver, spleen, kidneys and bladder. It is also beneficial for sciatica, spitting of blood and dysentery, helps jaundice and kills worms in children.

RIBWORT or
Ribwort Plantain
(Plantago angustifolia, Quinquenervia)

It grows in fields and meadows, and flowers in May and June.

It has a binding nature and has similar qualities as Common Plantain. It is particularly good for worms in children. The juice of it is given before the fit of an ague, to prevent its approaching.

RIE (Rye)
(Secale vulgatius)

It is of a hot and dry nature. Externally applied, ryemeal alleviates a headache, draws out thorns and splinters, and suppurates tumours and cysts. It can also be put into poultices for abscesses and gout.

ROCKET or Garden
White Rocket
(Eruca sativa alba)

It grows in gardens, and flowers in June.

It has a hot dry nature, and can be eaten with Lettuce, Purslain and other cold herbs in a salad. It excites sexual desire, especially the seed, which is also beneficial to the mouth.

WILD ROCKET
(Eruca sylvestris)

This is like the Garden Rocket, only the leaves are much smaller. It grows in stony places near highways and paths, and flowers most of the summer.

It has the same properties as Rocket.

WHITE ROSE
(Rosa alba)

It has a binding cooling nature. It is good against inflammations of the eyes, swellings of the breast and burning fevers. A water may be distilled from them which is very good for all eye troubles.

Take one handful of white roses, vervein, celandine, rue, fennel, eye-bright, daisy roots and houseleek, and boil them well in 2 quarts of Spring Water, until only 1 quart remains, then strain it, and add a noggin of pure virgin honey, boil it again, and then remove the scum that will form on the surface. Allow the remaining liquid to cool, and then bottle it. For some eye problems, such as webs or films, it may be necessary to add a little white Copperas to the liquid.

This remedy never fails to cure any external disorders of the eyes, even though all other cures may prove to be ineffectual.

DAMASK ROSE
(Rosa damascena et apllida)

The flowers aid digestion, purge bilious substances, opens obstructions of the liver and are good against jaundice. The distilled water is useful in fevers.

RED ROSE
(Rosa rubra)

The flowers stop diarrhoea, menstrual and other flows of blood, strengthen the stomach, prevent vomiting and tickling coughs, and are useful in consumption. A rose cake steeped in hot vinegar cures headaches, and pains from any inflammation. A conserve is beneficial for consumptive fevers.

ROSEMARY or
Narrow-Leaved Garden Rosemary
(Rosmarinus hortensis)

It is planted in gardens, and flowers in April and May.

It has a hot dry nature, and it is good against all disorders of the head and nerves. It secures loose teeth, and if rubbed with it, cleans them. It is useful for coughs, asthma, running nose and eyes, and obstructions of the lungs. Used in baths,

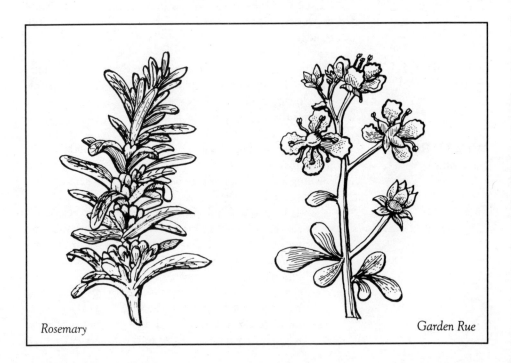

Rosemary

Garden Rue

it disperses clammy and viscous juice lodged in the joints and strengthens them. A fine powder made from one dram of the following cures migraine and apoplexy, Rosemary, Sage, Lily of the Valley, the tops of Sweet Marjoram, Nutmegs and Asarabacca roots, if sniffed into the nose.

GARDEN RUE
(Ruta hortensis)

It is good against infections and epidemical fevers. It cures colic pains and the bites of mad dogs. Made into a poultice with Garlic, Bay Salt and Bacon, it cures abscesses effectively; it assists in birth, expels the afterbirth and provokes menstruation. It is very good medicine for the eyes.

Take the leaves of Rue picked fresh from the stalks and bruise 6 oz, of Garlic, picked from the stalks, and bruised, Venice Treacle, and the scrapings of Pewter, 4 oz, boil all these over a slow fire, in 2 quarts of strong ale, until 1 pint of liquid has been consumed, then keep the remainder in a close stoppered bottle. Heat nine spoonfuls, and give seven mornings fasting.

This is an infallible cure for the bite of a mad dog if it is applied within nine days after a person has received the hurt.

WILD RUE
(Ruta sylvestris sive harmala)

It has three or four stalks growing upright, covered with long narrow leaves, smaller and more divided than Garden Rue. The flowers are white and appear on the tops of the branches.

It has a hot nature, therefore it disperses and dissolves viscous phlegm and provokes urination and menstruation.

RUPTURE WORT
or All Seed
(Millegrana minima)

It is a very small plant, having many small white flowers like little round grains. It is useful for ruptures, if it is applied in a poultice. Taken internally, it is good against colic pains.

COMMON CANDLE RUSH
(Juncus levis)

It has a dry nature. If the seed is parched and steeped in wine, it stops diarrhoea and menstrual flow and provokes urine.

S

·SAFFRON
(Crocus verus sativus)

It is cultivated in gardens.

It is very good for fevers, smallpox and measles, for jaundice and for most disorders of the lungs. Saffron also opens obstructions of the liver and spleen, provokes menstruation, assists birth and expels the afterbirth.

BASTARD SAFFRON or
Safflower
(Carthamus sive cenicus)

The stalks are round, growing to 3 or 4 ft high and it is covered with long indented prickly leaves. On the tops of the branches are several round prickly heads, from which appear pleasant orange-coloured flowers. It grows in gardens, flowering in July.

The juice of the seed drunk with honey and water purges viscous phlegm. It is good against coughs and the cholic and disorders of the lungs. The flowers open obstructions of the liver and are useful against jaundice.

COMMON GARDEN SAGE
(Salvia hortensis major vulgaris)

It has a very hot, dry and astringent nature. If eaten, it can benefit pregnant women, and can also promote conception. The juice drunk with honey is very useful against spitting or vomiting of blood. A decoction of water taken internally cures coughs, dysentery and obstructions of the liver and spleen. It is commonly used in gargles for sore mouths.

SMALL SAGE
(Salvia minor)

This Sage has much the same virtues of the former, only it is commonly chosen to make Sage tea, being accounted best for that purpose. It is a good antidote to love potions.

WOOD SAGE or
Garlic Sage
(Salvia agrestis vel scorodonia)

It has several square hairy brown stalks; its leaves are not much unlike the leaves of Great Sage, only that they are broader, softer and shorter. The flowers grow in long spikes on the tops of the branches. It is found in hedges and shrubs, and flowers in July.

It has a hot dry nature. A decoction of it drunk dissolves congealed blood and cures internal wounds and bruises. It is also useful in provoking urination and menstruation.

MOUNTAIN SAGE
(Salvia alpinea)

The leaves are rough, like Garden Sage and it grows in mountains.

It is very good at removing obstructions and aids conception, and is especially useful in provoking menstrual flow.

BLACK SALTWORT or Sea Miltwort
(Glaux, Martima exigua)

It grows in low salt marshes, and moist places near the sea, and flowers in June and July.

If eaten while nursing, it increases the production of milk. It has the same properties as Polygala or Milkwort.

SAMPHIRE
(Crithmum faeniculum marinum, Herba St. Petri)

It grows on rocks by the sea, and is especially abundant in the Aran island and the west coast of Clare.

It has a warm dry nature. It provokes urine and menstruation, opens obstructions of the bowels, liver, spleen and kidneys, strengthens the stomach, and creates an appetite if it is picked and eaten or a decoction drunk in wine.

Samphire

SANICLE
(Sanicula seu diapentia)

The leaves are dark green, five-cornered and serrated about the edges. The stalks are about 1 ft high, and are bare of leaves to the top on which appear little clusters of five-leaved white flowers. It grows in woods and on stony banks, and flowers in May.

It stops dysentery, and all other flows of blood. It heals ulcers of the kidneys, ruptures, decayed lungs and rotten sores of the mouth, gums and throat. It is also exceedingly good against internal and external wounds and bruises.

SARACENS CONSOUND
(Solidago saracenica)

The stalks are round and brown, and about 5 or 6 ft high. The leaves are long, narrow and indented about the edges and the flowers have a pale yellow colour. Saracens Consound flowers in August.

It has a dry astringent nature. It heals all kinds of wounds and ulcers, both internal and external. A decoction is very good against obstructions of the liver, bladder and gall. A gargle made from it is useful against ulcers of the mouth, throat and gums.

SAVIN or the Ordinary Savin tree
(Sabina)

It is planted in gardens, and has a very hot and dry nature. Drinking a decoction of the leaves in wine provokes urination and menstruation, causes abortion, and expels a stillborn child, and also the afterbirth. The crushed leaves applied with honey cures ulcers, scabby hands, spreading sores and warts. The juice of the leaves mixed with milk and sweetened with honey is a powerful remedy for destroying worms in children.

SUMMER SAVOURY
(Cunila hortensis aestiva)
WINTER SAVOURY
(Satureia hortensis vulgaris)

The winter variety grows wild and flowers in June.

Both varieties have a hot dry nature. They provoke menstrual flow, open obstructions, encourage flatulence, and are beneficial for the mouth, lungs and womb.

WHITE SAXIFRAGE or Stone Break
(Saxifraga alba)

It has whitish green round leaves, notched about the edges; the stalks are round and hairy growing about 1 ft high, on which grow spikes of white five-leaved flowers. The root consists of several small reddish grains, mixed with a few small fibres. It grows in dry meadows and rough stony places and flowers in April and May.

It has a hot dry nature. Taken internally, a decoction of the granulated root is very good against obstructions of the kidneys, bladder and ureter.

Savin

Scabius

MEADOW SAXIFRAGE
(Saxifraga vulgaris pratensis)

It has long channelled stalks over 2 ft high; the leaves are a dark green, and serrated about the edges, and the flowers are small and of a pale yellow colour. It grows in upland meadows and pastures, and flowers in August.

It is very good against obstructions of the kidneys, bladder and ureter. The crushed root taken with sugar aids digestion, eliminates flatulence and cures colic pains.

GOLDEN SAXIFRAGE
(Saxifrage aurea)

It grows to about a span and a half high; the leaves are like those of Meadow Saxifrage; on the tops of the stalks grow small golden flowers, which are succeeded by round husks full of small red seed. It grows in moist places, and flowers in March and April.

Its properties are similar to those of Meadow Saxifrage.

SCABIUS or Common Field Scabious
(Scabiosa major vulgaris)

When it first appears, the leaves are long, rough and hairy, spreading on the ground. The stalks are round and hairy, 2 or 3 ft high, bearing jagged leaves like the Great Valerian, and on the tops of which appear round, flattish blue flowers. It grows in meadows and pastures, and flowers in June and July.

It has a hot dry nature. Drinking a decoction of the leaves

clears the chest and lungs and is therefore good against coughs, abscesses and sore throats. It is good against scabby and itchy skin, if it is made into an ointment.

SCIATICA CRESSES
(Iberis)

It is above 1 ft in height, and bears spikes of small white four-leaved flowers. It grows in gardens, flowering in June.

An ointment made from it with Foxglove and Rue, cures sciatica. The leaves and root beaten into a plaster with Hogs lard has the same effect, but it must be kept on the afflicted part for only four hours.

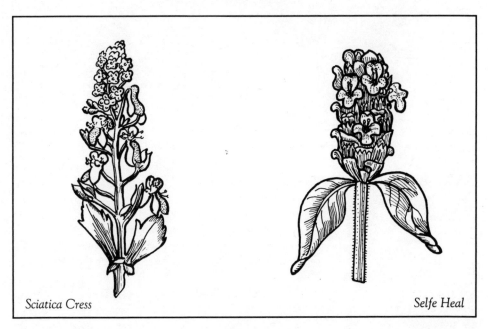

Sciatica Cress

Selfe Heal

SCORPION GRASS
or Mouse Ear
(Mysotis scorpioides arvensis)

It is a low herb, not much longer than a man's hand; the stalks are small, on which grow long narrow leaves, which are somewhat like a hare's ear and the flowers are small and yellow. It is planted in gardens, but often grows wild, and flowers in June and July. (There is also another kind called Water Scorpion Grass (Mysotis Scorpioides palustris), but its virtues are as Scorpion Grass.)

In properties, it is similar to Borage or Bugloss. The leaves and flowers infused in wine make the heart cheerful and merry. It is also good for the kidneys, and when nursing, helps increase the production of milk.

SCORZONERA or
Viper's Grass
*(Scorzonera seu vipravia
hispanica)*

It is planted in gardens, and flowers in July.

The root only is used. It works against poison and infection, induces perspiration and is good for the heart, and all kinds of fever.

**GARDEN SCURVY
GRASS**
*(Cochleria batava
rotundifolia hortensis)*

It grows in gardens, and flowers in April.

It cleanses and is exceedingly good to purify the blood, if the juice or an infusion of it is drunk. It is an excellent remedy for rotten ulcers of the mouth, if they are washed reguarly with a decoction. If it is applied with vinegar it cures scabs, itches, freckles and other eruptions of the skin.

**SEA SCURVY
GRASS**
(Cochleria marina)

It grows in salt marshes, and flowers later than the garden kind.

It has much the same properties as the Garden Scurvy Grass.

**WILD SCURVY
GRASS**
(Cochleria sylvestris)

It grows in fields and in shape and qualities it is similar to the Garden Scurvy Grass.

**COMMON BROAD
LEAVED SEA
WRACK** or Sea Weed
*(Quercus marina sucus,
Alga marina latifolia
vulgatissima)*

On the sea coast in the west of Ireland, the inhabitants dung or manure their land with it, and also burn it, in order to make ashes, which they transport to England, and there dispose of to good advantage as a fertilizer.

SEA THONGS
*(Quercus marina
secunda sucus longo
angustoq folio)*

These sea weeds are found upon our coasts, being cast up by the tide.

SELERY
*(Apium dulce,
Eleoselinum,
Paludapium)*

It has the same nature with Smallage and Parsley, and therefore partakes of their virtues and qualities.

SELFE HEAL
*(Prunella major folio non
dissecto)*

It grows in meadows and pastures, and flowers in June and July.

It dissolves clotted and congealed blood, heals all internal and external wounds, removes obstructions of the liver and gall, and is therefore good for jaundice. It cures ulcers of the mouth and gums, if they are washed with a decoction; if applied to the head with oil of roses and vinegar, it cures a headache. It is also good against internal bleeding and pissing of blood.

SENNA TREE or Bastard Senna (Colutea)

It grows in gardens.

The leaves and seed are violent purgatives, and should only be given to those with strong constitutions, and even then, they should be used with caution.

SHEPHERDS PURSE or Pickpurse or Caseweed (Bursa pastoris)

It grows on banks, in streets and backyards, and flowers all summer.

It has a very hot, dry and binding nature. It stops diarrhoea, dysentery, the spitting and pissing of blood and menstrual and other flows of blood. It is so excellent for this purpose, that it will stop bleeding, if it is only held in the hand. It is also very good in preventing miscarriages. The juice is useful against a fever.

SKIRRET (Sisarum)

It is planted in gardens, and flowers in June.

The roots have a hot and dry nature. They excite the appetite, provoke urine, stop diarrhoea, cure colic pains and also the hiccups.

SLAUKE or Sea Lettuce or Wrack or Laver (Lichen marinus, Muscus marinus, Fucus marinus, Lactuca marina)

It is a plant without stalks, having wrinkled broad leaves, like the leaves of lettuce, only more wrinkled. It grows on rocks in the sea.

It is good against boils and gout, and when boiled and dressed, it is esteemed by some as a delicate dish.

SLOE BUSH or Black Thorn (Prunus sylvestris)

Sloe trees grow in hedges and the fruit has a cold dry and astringent nature.

They stop diarrhoea, menstrual and all other flows of blood. It is used in gargles for sore mouths and gums and to fasten loose teeth. A decoction of the bark heals wounds and ulcers.

SMALLAGE (Apium palustre vel eleoselinum)

It grows in moist marshy places, and flowers in summer.

In qualities it is similar to Garden Parsley. It provokes urine, opens obstructions of the liver, spleen and kidneys, and encourages flatulence. The juice cleanses rotten sores, especially those of the throat and mouth. The leaves purify and sweeten the blood.

SNEEZEWORT or Common Field Pellitory
(Ptarmica vulgaris pratensis)

The stalks are upright, about 1 ft high; the leaves are long and narrow, finely serrated about the edges, while the flowers appear in clusters of white petals. It grows in moist meadows, and flowers in July.

When chewed, the root is good for toothache; if crushed and sniffed up the nostrils, it causes sneezing and clears the head.

SOLOMONS SEAL
(Polygonatum seu sigillum Solomonis)

It is planted in gardens, but grows wild in dry woods and copses, and flowers in May and June.

It has a hot dry nature, which is cleansing and somewhat binding. If pounded and applied to wounds, it closes and heals them. The juice of the root removes spots, freckles and black and blue marks caused by bruising. It stops haemorrhages and all kinds of flow.

Sopewort

SOPEWORT or Bruisewort
(Saponaria)

It grows in gardens, moist places and near rivers, and it flowers in June.

It has a hot dry nature. It clears the chest and lungs and is also good against epilepsy, and general infections. A decoction can be used to wash out wounds and rotten ulcers.

ORDINARY MEADOW SORREL
(Acetosa vel oxalis)

It grows in cornfields and meadows, and flowers in May.

The leaves are cooling, able to quench the thirst and resist putrefaction and are good in fevers.

FIELD SORREL or Sheep's Sorrel or Spear-Pointed Sorrel
(Acetosa arvensis)

It is like the Great Sorrel only much smaller. It grows in dry barren soil, flowering in May.

In qualities, it is similar to the Great Sorrel, but it is not so strong in its operation. It suppurates abscesses and tumours. Taken internally, it removes obstructions, is beneficial for jaundice and quenches the thirst.

ROUND LEAVED or **ROMAN SORREL**
(Acetosa romana rotundifolia hortensis)

It grows in gardens and has the same virtues and qualities to the Common Sorrel.

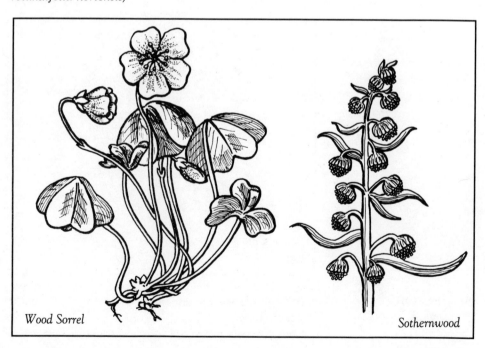

Wood Sorrel   *Sothernwood*

WOOD SORREL
(Acetosella lujula,
Trifolium acetosum
vulgare)

It grows in woody and shady places, and flowers in April.

It is more powerful than Common Sorrel. It is beneficial for the heart, mouth and liver and because it induces perspiration, it is also good against jaundice. It cleanses and heals rotten ulcers.

SOTHERNWOOD
or Male Common
Sothernwood
(Abrotanum mas)

It is planted in gardens, and flowers in July.

It has a very hot and dry nature. It provokes urination and menstruation, opens obstructions and breaks up stone in the bladder and kidneys, and kills worms. Applied externally, it is good against baldness and hair loss. It dissolves swellings, and can be put into an ointment that is generally strengthening.

PRICKLY SOW
THISTLE
(Sonchus asper)

It grows by banks and by waysides, and flowers in May and June.

It has a cold dry nature and it is of much the same nature with Dandelion. Is good against the pain of the stomach, and provokes urine, and breaks up stone in the bladder. If a decoction is drunk when nursing it helps to increase the production of milk. The leaves are useful against boils, abscesses and inflammations. The leaves roasted in the embers cure piles, and the falling down of the anus.

SMOOTH SOW
THISTLE or Hares
Lettuce
(Sonchus laevis)

This thistle grows in the same places with the former, and has the same virtues and qualities.

IVY LEAVED SOW
THISTLE or Wild
Lettuce
(Sonchus laevis muralis)

It is similar to Prickly Sow Thistle as to its medicinal operations.

SOW BREAD
(Artanita vel cyclamen)

The leaves are round, dark green on top and reddish or purple underneath, and usually marked with white spots. The flowers are pale purple, and the root is round like a small turnip. It is planted in gardens, but also grows wild, and it flowers in September and October.

It has a very hot dry nature. A dram crushed and taken, it opens obstructions of the liver and spleen, cures jaundice, and expels the afterbirth. The juice mixed with honey and dropped into the eyes clears the sight, and removes any impediment. It also cures mangy and itchy skin. The root hung about a women's neck in labour assists delivery. However, it is very dangerous for pregnant women to make use of it or to step over it.

SPIGNEL or Mew
(Meum vulgatius)

The leaves are not very large and are divided into several fine segments. The stalk is about 1 ft high, but without many branches, and on the tops of the branches grow clusters of small white five-leaved flowers. It flowers in June and only the root is used.

It has a hot dry nature. It opens obstructions of the liver, spleen and kidneys, provokes urine and eliminates flatulence from the stomach. It also cures stomach and colic pains and catarrh. Being applied to young children's bellies, it will cause them to make water plentifully.

SPINAGE
(Spinachia)

It has a dry moist nature. It provokes urine and aids digestion. Applied externally it dissolves inflammatory tumours and boils.

SPINDLE TREE or Common Prickwood with Red Berries
(Euonymus vulgatis)

It is a kind of shrub, rather than a tree, for it is small and low, not growing to any height. It grows in moist woods. It is thought to be hurtful to cattle and especially goats.

It is called Spindle tree and Prickwood because spindles and skewers are commonly made from the wood.

SPLEENWORT or Ceterach or Miltwast
(Asplenium vel scolopendria)

The leaves are almost the length of a man's finger, and $\frac{1}{2}$ in wide, green on top and brownish underneath. It has neither stalk nor flower. It grows in shady and stony places and on old stone buildings.

It has a temperate nature. The juice of the leaves taken in vinegar for forty days every morning, cures all obstructions of the spleen and liver. It is also very good against painful urination, stone in the bladder, jaundice, fevers and rickets in children.

ROUGH SPLEENWORT
(Lonchitis aspera minor)

The leaves are about 1 ft long, in shape like those of Polypodium, only much narrower and more finely divided. It grows along the sides of ditches, and in woody and low moist places.

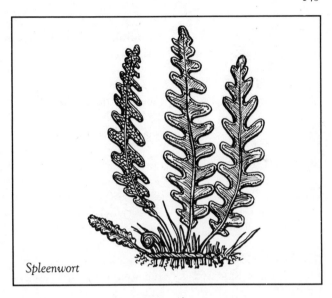

Spleenwort

It has a hot dry nature. Either internally or externally applied, it is very good for obstructions and swellings of the spleen. It is also beneficial for wounds, for it protects them from inflammation.

SPUNK or Touch Wood
(Fungus ignarus)

It grows on the stumps and trunks of trees in the shape of a horse's hoof.

SPUNGE
(Spongia)

This is an imperfect plant which grows on rocks and stone on the sea bed. It is never used internally in medicine, but can be used in applications externally to stop bleeding.

SPURGE or Laurel
(Laureola)

It grows in gardens, and flowers in March and April, with the berries ripening in September.

The leaves and the berries aid digestion and purge viscous substances. If the leaves are chewed, they draw out phlegmatic and clammy substances from the brain, and if sniffed into the nostrils, they cause sneezing. However, it is dangerous to take internally, for it can inflame the stomach.

PETTY SPURGE
(Esula rotunda sive paeplus)

This plant is fashioned like a small tree, like Sun Spurge, but far smaller.

It has a hot and dry nature, and it aids digestion, softens

hardness of the spleen, dissolves flatulence and expels viscous phlegm. It is dangerous in operation and should be used with great caution.

SEA SPURGE
(Tithymalus maritimus)

It produces six or seven red stalks. The leaves are like the leaves of flax, while the flowers are yellow, and are succeeded by triangular seed. The root is long and of a woody substance.

SUN SPURGE or
Wart Spurge
(Tithymalus helioscopium)

It has three or four stalks about 1 ft long which are reddish, the leaves are not as thick as garden Purslain, and the flowers are yellow and grow in tufts.

(It is called Helioscopium because it turns round with the sun.)

MOUNTAIN SPURGE or Knotted
Rooted Spurge
(Tithymalus hibernicus montanus)

All kinds of Spurge usually flower in June and July, with their seeds ripening in August, and they all have much the same properties.

They purge viscous phlegm, and the juice when put into hollow teeth, or the roots if boiled in vinegar and held in the mouth, alleviates the toothache. The juice also cures the roughness of the skin, itches, warts, decaying ulcers, and most eruption on the skin.

SPURRY
(Spergula)

It has round stalks, each having three or four joints, on which grow small narrow leaves in the shape of a star. On the tops of the stalks grow many small white flowers. It usually grows in fields and among corn. It is good fodder for cattle, for it increases their production of milk.

STONE BRAMBLE
or Raspis (The Juice of a Fair Woman)
(Rubus sanatilis, Alpinus, Chamerubus, Sanatilis, Rubus minimus)

It grows in moist woods, and bears red berries which are useful against scurvy.

SMALL STONE CROP or Wall Pepper
(Sedum parvum

It grows on old walls and buildings, and flowers in summer.

It has a hot dry nature. The juice taken with vinegar excites vomiting, and purges viscous phlegm. It is good

vermicularis,
Illecebra minor acris)

against fevers and poison, although it must be administered only to strong constitutions. The leaves mixed with lard dissolves lumps and swellings. If pounded and the juice drunk in beer, it is very good for fevers.

STRAWBERRIES
(Fragaria)

Drinking a decoction of strawberry leaves stops diarrhoea and menstrual flow and also cleanses the gums and cures sores and ulcers of the mouth. The juice of the leaves removes redness of the face. Strawberries quench the thirst, and are good for the heat of the stomach.

GARDEN SUCCORY
(Cichoreum sativum)

It is planted in gardens, and flowers in June.

It has a cold dry nature. Drinking the juice eases the pain of the stomach, provokes urine, breaks up stone and cures jaundice. When nursing, it increases the production of milk.

WILD BLUE SUCCORY
(Cichorium sylvestre)

It is like Garden Succory, only not so tall. It grows in lanes and by hedgesides, and flowers in July.

It has the same properties as the former. It opens obstructions and is good against scurvy.

SUN DEW
(Ros folis rota folis vel sponsa solis)

It has reddish rough round leaves about an inch long, shaped like spoons, growing on long stalks, bearing several

Sun Dew

small five-leaved flowers. It grows in bogs and moist bottoms and flowers in June and July. It has a strange nature, for the hotter the sun shines on it, the moister it is.

If bruised with salt and laid on the skin, it raises blisters. Being corrosive, it eats away at rotten sores.

SWALLOW WORT
(Asclepias vincetoxicum, Hirundinaria)

The stalks are smooth, round and small and about 2 ft high. The leaves are dark green, like ivy leaves, only longer and sharp pointed; on the tops of the stalks grow small bunches of five-leaved white flowers. It grows in gardens, and often in sandy and rocky mountainsides, and flowers in June.

The roots have a hot and dry nature, and are a useful antidote against poison. Drinking a decoction cures colic pains, and is good for the bite of a mad dog, used either internally or externally. Pounded leaves are good in applications to sore breasts. The roots are good in inducing perspiration and are useful against jaundice.

T

TAMARISK
(Tamariscus)

It is a small tree that grows in gardens.

The wood, bark and leaves are very good for all disorders of the spleen. Drinking a decoction opens obstructions and is good for coughs and catarrh.

COMMON YELLOW TANSY
(Tanacetum)

It grows near highways, and along the borders of fields, and flowers in July.

It has a hot dry nature, the seed and flowers are an approved remedy against worms. Bruised and mixed with Ox Gaul and applied to the region of the navel, it kills worms in children. Rubbing a mixture of Oil of Roses and the juice of Tansy to the body prevents a shivering fit. The juice drunk with wine provokes urine and opens all obstructions of the body. An ointment made of it cures scabby and itchy skin.

Common Yellow Tansy

WILD TANSY or
Silver Weed
(Argentina vel potentilla)

It grows in moist barren places, and flowers in May and June.

Drinking a decoction is good against diarrhoea, dysentery, menstrual and other flows of blood. Taking a decoction in water and salt is good against bruising. It cures ulcers of the mouth and gums, heals wounds, secures loose teeth, eases a toothache and removes freckles, pimples and sun burn.

STRANGLE TARE
or Wild Vetch
(Aracus seu cracca major)

The leaves, stalks and pods are like the Common Vetch, only much smaller. The leaves are Cloven at the ends into two or three clasping tendrils. The flowers are small and of a light purple colour, and the pods are long and narrow.

There is little of medicinal value in it, but cattle feed upon it.

MANURED TARE
or Vetch
(Vicia sativa)

It is sown in fields, and flowers in May, with the seed ripening in August and September.

A decoction in milk drives out smallpox and measles, but in general it is not much used in medicine.

TARRAGON
(Dracunculus hortensis seu draco herba)

It is planted in gardens, and flowers in July and August.

It has a hot dry nature. It provokes urination and menstruation, and expels flatulence.

GARDEN TEASEL
or Manured Teasel
(Dipsacus sativus seu carduus sullonum)

The leaves are long and large, serrated about the edges; the stalks are about 3 ft high, divided into several branches, on the tops of which appear large heads, full of crooked pricklyhooks. It is cultivated in fields, flowering in July.

WILD TEASEL or
Great Shepherds Rod
or Venus Baton or
Carde Thistle
(Dipsacus syvestris, Labrum venaris, Virga pastoris major)

It is like the former, only the leaves are narrower, and it has purple flowers, and its hooks are neither so hard nor so sharp. It grows in moist places, near brooks and rivers, and flowers in July.

The roots boiled in wine and pounded into a smooth ointment heal anal ulcers, and take away warts, although this ointment must be retained in a copper box. The water

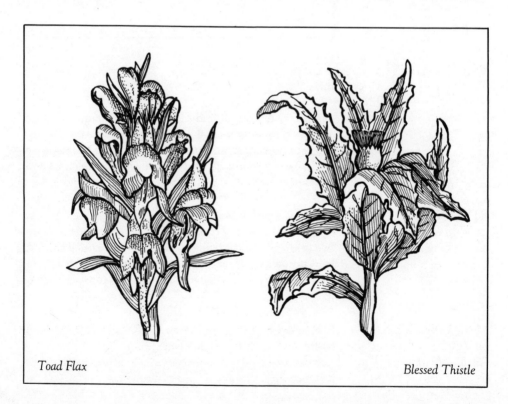

Toad Flax

Blessed Thistle

which is found in the cavities of the leaves is good for the eyes and face.

BLESSED THISTLE
(Carduus benedictus)

It grows in gardens, and flowers in June.

It has a hot and dry nature, and it provokes urination and menstruation, induces sweat, and is beneficial for the mouth, lungs and kidneys. Drinking a decoction in wine cures colic pains. In its powdered state it is exceedingly good against pestilence, but it must be taken within the space of twenty-four hours from the time a person is seized with it. The juice is a powerful antidote against poison if it is taken immediately. Drinking a decoction with water provokes a gentle vomit. It also destroys worms in the stomach.

LADY'S THISTLE
or Milk Thistle
(Carduus mariae)

The leaves are long, large, prickly and shiny green. The stalks are as thick as a man's finger, growing to 5 ft in height, and on the tops of the stalks appear large heads full of sharp prickles. It grows in uncultivated places on banks and along the borders of fields, and flowers in June.

The root is dry and astringent, and the seed is hot. Drinking a decoction of the root in wine is good for the spitting of blood and for weak stomachs. The seed may be profitably given to children who suffer from convulsions.

GARDEN THYME
(Thymus)

It is planted in gardens, and flowers in July.

It has a hot dry nature. It is good against coughs, shortness of breath and all disorders of the head and nerves. Two drams of its powder in vinegar and honey with a little salt, clears phlegm. If pounded with vinegar it dissolves swellings and removes warts. If pounded with barley meal and applied it is good for sciatica.

TOAD FLAX or
Great Toad Flax
(Linaria)

It has small slender blackish stalks; the leaves are long and narrow with sharp points, and have a bluish-green colour, and the flowers are yellow. It grows in uncultivated places and along the borders of fields, and flowers in July.

It has a hot dry nature. It provokes urine, and opens obstructions of the liver spleen and kidneys. It also breaks up stone in the bladder and is useful for jaundice. An ointment made with hogs lard mixed with an egg yolk is excellent for piles.

TOBACCO
(Nicotiana,
Petum,
Tabaccum,
Hyoscyamus peruvianus)

It grows in gardens, and flowers in July and August.

It is good for clearing the head and glands of the throat of phlegm, so preventing catarrh, apoplexy, migraine, inflammation of the throat and infection. It encourages sneezing and vomiting, and it's a good antidote to poison. Externally applied, it cures an itchy and scabby skin.

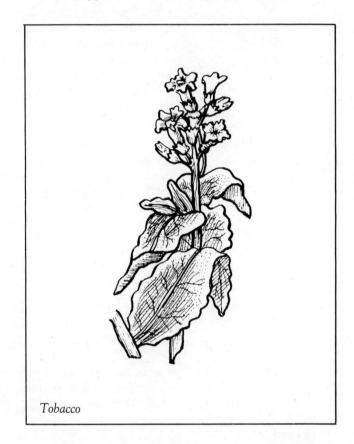

Tobacco

TORMENTIL or
Setfoil
(Tormentilla sylvestris)

It grows in moist meadows and commons, and flowers in June and July.

The root only is used in medicine. If crushed and drunk in wine or water, in which iron or steel has been quenched, it induces perspiration. It stops diarrhoea, dysentery, the spitting and pissing of blood, and menstrual and other flows of blood. It cures ulcers and other sores of the mouth if they are washed out with a decoction.

COMMON PURPLE MEADOW TREFOIL & WHITE FLOWERED MEADOW TREFOIL (Shamrock)
(Trifolium pratense purpureum minus, Trifolium pratense album)

They grow in fields and meadows, and flower in May and June.

They have a binding nature, good for all kinds of flow and for painful urination. Externally applied they are good against abscesses and inflammations.

TURNEP or Turnip
(Rapum)

It grows in gardens, and flowers in April.

It is a very wholesome nourishing root and is good for the chest, being useful against coughs and consumption.

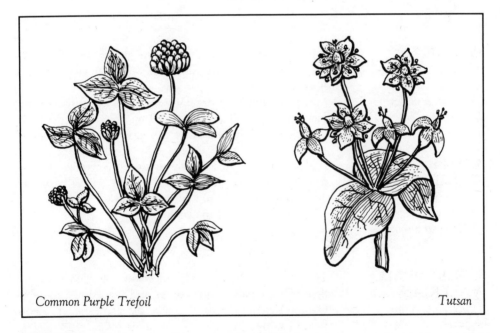

Common Purple Trefoil

Tutsan

TUTSAN or Park Leaves
(Androsaemum vulgare seu hypericum maximum)

It is very like St John's wort, except that its leaves and flowers are larger. It grows in hedges and thickets, and flowers in July.

This can be mixed with other medicines in treating

wounds, and is also good for the kidneys. Otherwise its properties are similar to those of St John's Wort.

TWAYBLADE
(Bifolium sylvestre)

It has a round smooth stalk which bears only two leaves, which are like the leaves of great Plantain. The flowers grow in spikes which have a dull green colour. It grows in thickets and moist meadows, and flowers in June.

It is good for wounds and ruptures.

V

GARDEN VALERIAN or
Setwall
(Valeriana major hortensis)

It has broad oval leaves of a whitish-green colour; the stalks are knotty and channelled and on them grow whitish flowers. It is planted in gardens.

The root has a hot dry nature; it provokes urination and menstruation and induces sweating. It is an antidote to poison and it is beneficial for the nerves, head and mouth.

LESSER VALERIAN or
Ladder to Heaven or
Jacobs Ladder
(Valentina minor)

It is like the garden variety only that it is smaller and the flowers have a pale purple colour. It grows in marshy ground and moist meadows, and flowers in May.

It is seldom used in medicine.

GREAT WILD VALERIAN
(Valeriana major sylvestris)

The stalks are channelled and about 3 ft high, and the flowers are a pale purple colour and shaped like those of Garden Valerian. It grows in moist places, near ditches, and flowers in May.

It heals ulcers and blisters of the mouth and inflammations of the throat, if the mouth and throat are washed with a gargle. It is good for the head and nerves. The roots being pulverized and the quantity of half a spoonful taken at a time in a pleasing drink is exceeding good against epilepsy.

A decoction drunk every day for a month has the same effect.

VERVEIN or Common Blue-Flowered Vervein
(Verbena communis flore caeruleo)

It grows in gardens, and is often found growing wild near walls, ditches and highways, flowering in July.

It is good for ulcers of the mouth and jaws if the mouth is washed with a decoction. Such a decoction will also secure loose teeth and alleviate the pain of a toothache if the liquid is held in the mouth for a while. If it is mixed with oil of roses and vinegar, it cures a headache. If the leaves are pounded with honey, they cure and heal fresh wounds. It is very good for all disorders of the brain and is excellent for strengthening the sight.

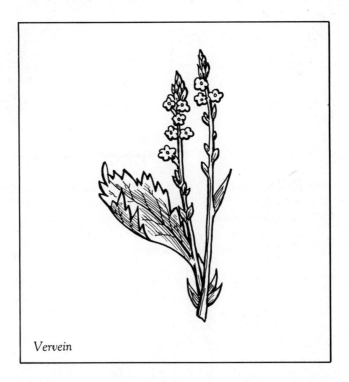

Vervein

VINE TREE
(Vitis vinisera)

In this country it is planted in some gardens, but it seldom comes to any great perfection.

The wine produced from the fruit, especially the red wine, aids digestion, strengthens the stomach and bowels and is good for the heart and is useful against infection.

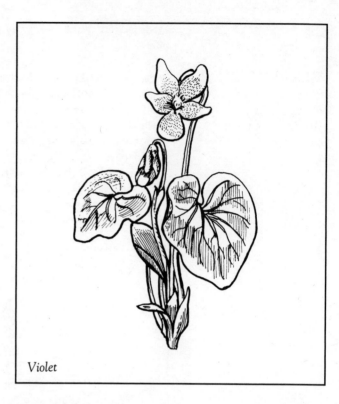

Violet

MARCH VIOLET
or Purple Violet
(Viola martina purpurea)

It has a cold moist nature. A decoction or syrup of it is good against fever, inflammations of the liver and lungs, and is useful for coughs and pleurisy. Externally applied, it is good for abscesses and inflammations. The seed breaks up kidney stone.

W

WALL FLOWER or Ordinary Yellow Wallflower
*(Keiri,
Cheiri,
Leucorum luteum
vulgare)*

It is planted in gardens and it is found growing wild on old walls and buildings, flowering in April and May.

It has a hot dry nature. It provokes urination and menstruation and expels a stillborn child, and the afterbirth if a decoction of the dried flowers , or a little seed is drunk in wine. A plaster made from the flowers with oil and wax closes up old ulcers. The juice dropped into the eyes strengthens the sight. It cures sores on the fingers and abscesses and is also good against apoplexy.

WALNUT TREE
(Juglans)

It is planted along walks, and in parks and fields.

Two or three walnuts eaten with a fig and a little Rue on an empty stomach, provide a good prevention against infection. The kernel oil will heal bruises and scabby and itchy skin, and taken internally will break up stone in the bladder and urinary crystals. A decoction of the green peel or husk of the walnut is useful against tumours and ulcers of the mouth and throat. The bark of the tree, either green, dried or crushed, encourages vomiting.

WHEAT
(Triticum)

A decoction of the flowers of wheat and honey and water dissolves tumours. A decoction of the bran in vinegar is good against itchy skin and boils. A piece of toasted bread dipped in wine and applied to the stomach stops vomiting. A poultice of the wheaten bread in milk will reduce tumours and a decoction of the flowers in vinegar and honey removes spots and freckles. It is advisable not to eat the crust of the bread especially if it is in the least bit burnt.

WILLOW TREE
*(Salix vulgaris alba
arborescens)*

A decoction of the leaves, bark, flowers and seed in wine, taken internally, stops vomiting, spitting of blood, excessive menstrual flow and all other flows of blood. The ashes of the bark mixed with vinegar causes warts to fall off, and softens hard skin. The sap that flows from the bark is good for inflammations of the eyes.

RUE LEAVED WHITLOW GRASS
(Paronychia rutaceo folio)

The leaves are thick, fat and divided into three parts; the stalks are hairy, and on their tops grow small, white five-leaved flowers. It is a small plant, about 3 or 4 in high. It grows on the tops of walls and low houses, flowering in April

It is beneficial against scrofulous tumours.

RED WINTER CHERRIES or Red Nightshade
(Alkekengi seu halicacabum)

It is planted in gardens, and grows wild in woods and moist places, and flowers in July and August, with the fruit ripening in September.

The leaves are cooling as with Common Nightshade. The berries provoke urine, and open obstructions of the bladder, liver and kidneys.

WINTER GREEN
(Pyrola vulgaris)

It has nine or ten green tender leaves, like those of Beet, but much smaller. The stalks are about 1 ft high, and on their tops appear several five-leaved white flowers. It grows in woods and shady places, and flowers in July.

The leaves are useful in healing internal and external wounds and ulcers. They stop haemorrhages, like the passing of blood in the urine and excessive menstrual flow.

WOAD
(Glasium sativum aut isatis sativa)

It is planted in fields and gardens, mainly as a principal ingredient in blue dyes. It is also found growing wild in uncultivated areas.

It is good when applied to wounds and decaying ulcers. It prevents bleeding and dissolves swellings. Drinking a decoction is useful against obstructions of the liver and spleen. It can also be applied to ruptures and dislocations.

WOLFE-BANE
(Aconitum)

It grows by riversides and near bogs and woods.

If made into a plaster and externally applied, it is very good in dissolving glandular tumours.

WOODROOF or Small, Sweet Scented Madder
(Asperula, Aspergula oderata, Rubeola montana odera)

The stalks are square and full of joints, at each of which grow seven or eight long narrow leaves, like a star. The flowers grow on the tops of the stalks in small clusters of little single-leaved white flowers. It grows in woods and shady places, and flowers in May.

It has a hot dry nature. It is good in healing wounds if

Wolfe-bane

bruised and then applied and also in curing boils and inflammations. If it is drunk with wine it is good for the heart and useful against inflammations of the liver and obstructions of the gall and bladder.

COMMON WORMWOOD
(Absinthium latifolium vulgare)

It strengthens the stomach, creates a good appetite, and can be useful for jaundice. It is good in stopping diarrhoea and vomiting. If applied externally, it dissolves swellings, and is good for dislocations, swelling of the tonsils and inflammation of the throat if a plaster is made from it with Rue, Sothernwood and hogs lard is applied.

SEA-WORMWOOD
(Absinthium tenuifolium vel seriphium marinum album)

If a decoction is drunk, or if applied externally to the belly or navel in a poultice it kills all kinds of worms.

Y

YARROW or Milfoil
(Millefolium terrestre vulgare)

It commonly grows in fields and meadows, flowering in June and July.

It has a very dry astringent nature. Drinking a decoction stops dysentery, and excessive menstrual and other flows. If bruised and applied to wounds, it stops bleeding and prevents inflammations and swelling. A dram of it pulverized and taken in a glass of white wine is a perfect remedy for the cholic. Nothing is more effectual against the piles, either taken inwardly or outwardly applied. If applied to the pit of the stomach in a plaster with grated nutmeg, it is beneficial for fevers.

YEW TREE
(Taxus)

It is good neither as food nor as medicine, and can actually be dangerous to man. It is reported that if a man sleeps under the shadow of this tree, he will fall sick and will sometimes die. Birds that eat of the fruit either die or cast their feathers. Eating the fruit can provoke acute diarrhoea.

Of further interest . . .

DICTIONARY OF MODERN HERBALISM
THE COMPLETE GUIDE TO HERBS

Simon Mills MNIMH. Herbal medicine is a healing art that has been practiced for thousands of years. In China it is held in equal esteem with acupuncture and Western medicine. Now, Simon Mills, the Director of Research and a past President of the National Institute of Medical Herbalists, has drawn upon his years of experience in herbal research to provide an up-to-date, concise reference book which is invaluable to both the general and professional reader.

It contains the latest information on the full range of herbs and herbal remedies and includes colour plates to assist in identification.

This title is being published by Thorsons in collaboration with Newman Turner Publications, publishers of HERE'S HEALTH magazine.

THE HEALING POWER OF HERBAL TEAS

Ceres. Colour photographs and line drawings. Lists common herbs, tells where and when to gather them and beautifully illustrates each one. Reveals in simple language the enormous potential of herbal teas as natural healers and tonics. Many of these easily obtained plants provide helpful alleviators of pain and simple discomforts where habit-forming drugs might previously have been used. Gives precise instructions for making infusions to relieve indigestion, cramp, insomnia, and sunburn, as an aid to slimming and as a general bodily tonic. Also gives directions for growing herbs either in an outdoor garden or an easily maintained window box, and for companion planting with non-edible species to enhance the appearance of the plot. A veritable cornucopia of refreshing and soothing drinks and lotions. A Thorsons publication.